FIBER OPTIC REFERENCE GUIDE

A Practical Guide to Communications Technology
Third Edition

by David R. Goff,
President & CTO
Force, Incorporated

Edited by
Kimberly Hansen

Assistant Editor

Michelle K. Stull

Focal Press
Taylor & Francis Group

NEW YORK AND LONDON

First published 2002

This edition published 2014
by Focal Press
70 Blanchard Road, Suite 402, Burlington, MA 01803

Simultaneously published in the UK
by Focal Press
2 Park Square, Milton Park, Abingdon, Oxon OX14 4RN

Focal Press is an imprint of the Taylor & Francis Group, an informa business

Notices
Practitioners and researchers must always rely on their own experience and knowledge in evaluating and using any information, methods, compounds, or experiments described herein. In using such information or methods they should be mindful of their own safety and the safety of others, including parties for whom they have a professional responsibility.

To the fullest extent of the law, neither the Publisher nor the authors, contributors, or editors, assume any liability for any injury and/or damage to persons or property as a matter of products liability, negligence or otherwise, or from any use or operation of any methods, products, instructions, or ideas contained in the material herein.

ISBN 13: 978-0-240-80486-6 (pbk)

Library of Congress Cataloging-in-Publication Data
Goff, David R.
    Fiber optic reference guide : a practical guide to communications technology / by David R. Goff; edited by Kimberly Hansen; assistant editor Michelle K. Stull.—3rd ed.
        p.cm.
    Includes bibliographical references and index.
    ISBN 0-240-80486-4 (alk. Paper)
    1. Fiber optics. 2. Optical communications. I. Hansen, Kimberly S. II. Title

    TA1800 .G64 2002
    621.382'75—dc21

                                                                2002018828

British Library Cataloguing-in-Publication Data
A catalogue record for this book is available from the British Library.

CONTENTS

Contents

Contents

Contents

FOREWORD

Over the last decade, fiber optics has emerged as a key technology component for today's information-centric society. Fiber optics has overcome numerous obstacles along the way. Early on, fiber optics suffered from the "chicken or egg first" syndrome. In other words, the technology would be available at a reasonable price once the demand for the technology was demonstrated. However, the demand would not arise until affordable technology was in place. The second obstacle to growth of the fiber optics industry has been standards. In the early years, the industry suffered from too few standards. Lately, the industry suffers from too many standards that are nearly obsolete by the time they are introduced. Last, competition from other technologies has been intense. No one would have guessed ten years ago that video would be sent over twisted pair copper cable for substantial distances, yet today, it can be done.

The fiber optics community has overcome these obstacles, which plague nearly all new technologies, to make fiber optics a viable, keystone technology for the Communications Age. Today, the fiber optics industry is at the dawn of a new era, the multi-terabit all-optical age. Real systems are now being deployed at this writing with a capacity of one terabit (10^{12} bits) per second per fiber. With society's exponentially increasing demand for data transport, the key question is whether the second age of fiber optics will have the required capacity. Already, next generation systems with a capacity five to ten terabits (5-10 x 10^{12} bits) per second per fiber are being designed and planned.

ABOUT THIS BOOK

In many technical areas, a reference book such as this one could go five or ten years before a third edition is warranted. Fiber optics in the early in the twenty-first century is not one of those areas. Several years ago, it looked like fiber optics was set for many years of steady, but not spectacular, growth. Then came the video revolution and the digital revolution, and the Internet came of age. Less than two decades ago, many people wondered why anyone would need a personal computer at home; today, the question of why individuals would need access to tens of megabytes of data per hour has been resoundingly answered by e-mail and the world wide web. Phrases like "paradigm shift" and "dawn of a new era" seem cliché, but today, no one doubts their applicability to the Internet and its impact on the telecommunications industry and society as a whole. Where telecommunications used to be about voice traffic, it is now, and forevermore, about data — exponentially growing amounts of data.

Fiber optics has been revised to live up to its decades old promise of terabit per second data rates. This third edition of the *Fiber Optic Reference Guide* describes all of the new technology that makes these stunning data rates possible. It also describes the many gremlins that hide along the way. The first edition of this reference guide started out as an educational tool for my company's customers. While the information contained herein has been greatly expanded, the original spirit of this book remains. The intent is still a very practical look at the technology and industry. In a sense, it is an attempt to present a very technical subject in the most straightforward and intuitive manner possible. Because of this approach, the text is not cluttered with lots of esoteric equations. The equations that are presented are those necessary to deal with fiber optic technology in real applications. There are many good, deep theoretical texts on fiber optics. You will recognize them because they will tell you, for instance, that Refractive Index (η) is identical to the square root of the relative dielectric constant (ε_r). Personally, explanations like that do not help me much. To me, it seems sufficient to view refractive index as a property of a material that causes it to bend light and causes light to travel proportionally slower compared to its speed in a vacuum. That is something that I can relate to and can use in calculations about real systems. This book will appeal to those who are newcomers to the field or those who cannot get past the techno-jargon that seems to go with every technology. The *Fiber Optic Reference Guide* presents the essential concepts of the fiber optics industry and gives the reader a good feeling of how the technology really works rather than presenting endless pages of mind-numbing equations. Insights into the history of fiber optics and its components are also included in this book. I think this is important because it helps clarify why things are done the way they are. By having some sense of history, what worked and what did not, it becomes easier to predict what the future will bring.

ACKNOWLEDGMENTS

The content of this book continues to evolve and has benefited from the input of a great many people to whom I am thankful. I would especially like to thank the editor, Kimberly Hansen, whose skill as a writer, graphic designer and advanced desktop publisher helped make this book possible. In addition to the knowledge of Force, Incorporated engineers and technical writing staff, input and information was received from a number of companies and individuals. I also thank all of those who contributed to make this a useful reference guide for the fiber optics industry and its present and future customers, including:

Amphenol, Incorporated; Corning Incorporated; Don Davis, Bandwidth9; Hewlett-Packard; Highwave Optical Technologies; Hopecom, Inc.; Dr. Ira Jacobs, VPI & SU; Melcor Corporation; Micron Optics, Inc.; Nortel Networks Optical Components; Optical Cable Corporation; PD-LD, Incorporated; Telcordia.

David R. Goff

President & CTO
Force, Incorporated
825 Park Street
Christiansburg, VA 24073
TEL: (540) 382-0462
E-mail: dgoff@forceinc.com

A HISTORY OF FIBER OPTIC TECHNOLOGY

1

THE NINETEENTH CENTURY

In 1870, the British Royal Society in London, England witnessed a thought-provoking demonstration given by natural philosopher, John Tyndall. Tyndall, using a jet of water that flowed from one container to another and a beam of light, demonstrated that light used internal reflection to follow a specific path. As water poured out through the spout of the first container, Tyndall directed a beam of sunlight at the path of the water. The light, as seen by the audience, followed a zigzag path inside the curved path of the water. This simple experiment, illustrated in Figure 1.1, marked the first research into the guided transmission of light.

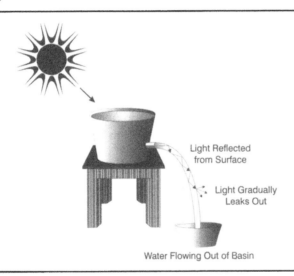

Figure 1.1: John Tyndall's Experiment

Light Reflected from Surface

Light Gradually Leaks Out

Water Flowing Out of Basin

William Wheeling expanded upon Tyndall's experiment when, in 1880, he patented a method of light transfer he called "piping light." Wheeling believed that by using mirrored pipes branching off from a single source of illumination, i.e. a bright electric arc, he could send the light to many different rooms in the same way that water, through plumbing, is carried throughout buildings today. Due to the ineffectiveness of Wheeling's idea and to the concurrent introduction of Edison's highly successful incandescent light bulb, the concept of piping light never took off until late in the 20th century when several commercial systems were introduced. Wheeling's ideas were about a century ahead of the technology required to make "piping light" feasible.

That same year, Alexander Graham Bell developed an optical voice transmission system he called a photophone. Like Wheeling's method of light transfer, the photophone preceded by decades the technology required to make it a commercially viable idea. The photophone used free-space light to carry the human voice 200 meters. Specially placed mirrors reflected sunlight onto a diaphragm attached within the mouthpiece of the photophone. At the other end, mounted within a parabolic reflector, was a light-sensitive selenium resistor. This resistor was connected to a battery that was, in turn, wired to a telephone receiver. As one spoke into the photophone, the illuminated diaphragm vibrated, casting various intensities of light onto the selenium resistor. The changing intensity of light altered the current that passed through the telephone receiver which then converted the light back into speech. The technology to support this invention would not be available for many years, but Bell believed this invention was superior to the telephone because it did not need wires to connect the transmitter and receiver. It was, in fact, the world's first optical amplitude modulation (AM) audio link. Today, "free-space" optical links, similar in concept to Edison's photophone, find extensive use in metropolitan applications.

THE TWENTIETH CENTURY

Fiber optic technology experienced a phenomenal rate of progress in the second half of the twentieth century. Early success came during the 1950's with the development of the fiber-scope. This image-transmitting device, which used the first practical all-glass fiber, was concurrently devised by Brian O'Brien at the American Optical Company and Narinder Kapany and colleagues at the Imperial College of Science and Technology in London. (In fact, Narinder Kapany first coined the term "fiber optics" in 1956.) Early all-glass fibers experienced excessive optical loss, the loss of the light signal as it traveled the fiber, limiting transmission distances.

This motivated scientists to develop glass fibers that included a separate glass coating. The innermost region of the fiber, or core, was used to transmit the light, while the glass coating, or cladding, prevented the light from leaking out of the core by reflecting the light within the boundaries of the core. This concept is explained by Snell's Law which states that the angle at which light is reflected is dependent on the refractive indices of the two materials — in this case, the core and the cladding. The lower refractive index of the cladding (with respect to the core) causes the light to be angled back into the core as illustrated in Figure 1.2.

Figure 1.2: Optical Fiber with Cladding

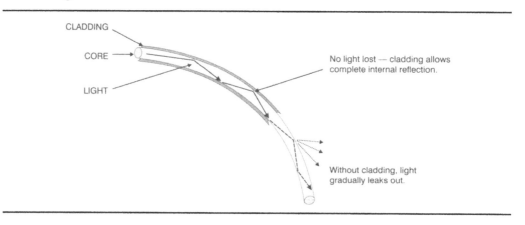

The fiberscope quickly found application inspecting welds inside reactor vessels and combustion chambers of jet aircraft engines as well as in the medical field. Fiberscope technology has evolved over the years to make laparoscopic surgery one of the great medical advances of the twentieth century.

The development of laser technology was the next important step in the establishment of the industry of fiber optics. Only the laser diode (LD) or its lower-power cousin, the light-emitting diode (LED), had the potential to generate large amounts of light in a spot tiny enough to be useful for fiber optics. In 1957, Gordon Gould popularized the idea of using lasers when, as a graduate student at Columbia University, he described the laser as an intense light source. Shortly after, Charles Townes and Arthur Schawlow at Bell Laboratories supported the laser in scientific circles. Lasers went through several generations including the development of the ruby laser and the helium-neon laser in 1960. Semiconductor lasers were first realized in 1962; these lasers are the type most widely used in fiber optics today.

Because of their higher modulation frequency capability, the importance of lasers as a means of carrying information did not go unnoticed by communications engineers. Light has an information-carrying capacity 10,000 times that of the highest radio frequencies being used. However, the laser is unsuited for open-air transmission because it is adversely affected by environmental conditions such as rain, snow, hail, and smog. Faced with the challenge of finding a transmission medium other than air, Charles Kao and Charles Hockham, working at the Standard Telecommunication Laboratory in England in 1966, published a landmark paper proposing that optical fiber might be a suitable transmission medium if its attenuation (loss of signal strength) could be kept under 20 decibels per kilometer (dB/km). At the time of this proposal, optical fibers exhibited losses of 1,000 dB/km or more. At a loss of only 20 dB/km, 99% of the light would be lost over only 3,300 feet. In other words, only 1/100th of the optical power that was transmitted reached the receiver. Intuitively, researchers postulated that the current, higher optical losses were the result of impurities in the glass and not the glass itself. An optical loss of 20 dB/km was within the capability of the electronics and opto-electronic components of the day.

Intrigued by Kao and Hockham's proposal, glass researchers began to work on the problem of purifying glass. In 1970, Drs. Robert Maurer, Donald Keck, and Peter Schultz of Corning succeeded in developing a glass fiber that exhibited attenuation at less than 20 dB/km, the threshold for making fiber optics a viable technology. It was the purest glass ever made. Concurrent advances in laser technology, semiconductor chips, optical detectors, and optical connectors, combined with the optical fiber, ushered in the true beginnings of the fiber optic communications industry.

The early work on fiber optic light sources and detectors was slow and often had to borrow technology developed for other reasons. For example, the first fiber optic light sources were derived from visible indicator LED's. As demand grew, light sources were developed for fiber optics that offered higher switching speed, more appropriate wavelengths, and higher output power.

Fiber optics developed over the years in a series of generations that can be closely tied to wavelength. Figure 1.3 shows three curves. The top, dashed, curve corresponds to early 1980's fiber, the middle, dotted, curve corresponds to late 1980's fiber, and the bottom, solid, curve corresponds to modern optical fiber. The earliest fiber optic systems were developed at an operating wavelength of about 850 nm. This wavelength corresponds to the so-called "first window" in a silica-based optical fiber. This window refers to a wavelength region that offers low optical loss. It sits between several large absorption peaks caused primarily by moisture in the fiber and Rayleigh scattering, the scattering of light particles due to impurities in the glass, at shorter wavelengths.

Figure 1.3: Four Wavelength Regions of Optical Fiber

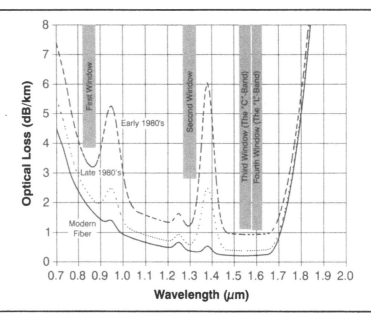

The 850 nm region was initially attractive because the technology for light emitters at this wavelength had already been perfected in visible indicator LED's. Low-cost silicon detectors could also be used at the 850 nm wavelength. As technology progressed, the first window became less attractive because of its relatively high 3 dB/km loss limit.

Most companies jumped to the "second window" at 1310 nm, but a few companies, notably IT&T, spent effort developing the wavelength region near 1060 nm. The 1060 nm region allowed low-cost silicon detectors to be used; however, the light emitter technology was more complex than 850 nm. This region did offer lower attenuation, about 1.7 dB/km. Ultimately however, the second window at 1310 nm won out with lower attenuation of about 0.5 dB/km. In late 1977, Nippon Telegraph and Telephone (NTT) developed the "third window" at 1550 nm. It offered the theoretical minimum optical loss for silica-based fibers, about 0.2 dB/km.

Today, 850 nm, 1310 nm, and 1550 nm systems are all manufactured and deployed along with very low-end, short distance, systems using visible wavelengths near 660 nm. Each wavelength has its advantage. Longer wavelengths offer higher performance, but always come with higher cost. The shortest link lengths can be handled with wavelengths of 660 nm or 850 nm. The longest link lengths require 1550 nm wavelength systems. A "fourth window," near 1625 nm, is being rapidly developed. While it is not lower loss than the 1550 nm window, the loss is comparable, and it might simplify some of the complexities of long-length, multiple-wavelength communications systems.

Most optical fiber is based on silicon, like most of today's electronics. There have been numerous attempts over the last few decades to develop alternate materials to either reduce cost or improve performance. To date, none have shown any real promise in dethroning silicon-based fiber. Much effort has gone into developing plastic fiber, but its impact on the marketplace has been minimal. It is suited for very short distances, typically around tens of meters only. As the cost of glass fiber has plunged over the last decade, the advantage of plastic fiber has faded.

The other big push has been to develop heavy metal halide fibers for use at long wavelengths in the 3 μm to 5 μm region (two to three times longer than the third window). The attraction here is that, in theory, such a fiber could have optical losses as low as 0.001

dB/km. However, after much effort, the losses of such fibers cannot even match the best losses of silica-based fibers. It seems likely that silicon will continue to be the basis of most of today's high technology.

Real World Applications

The U.S. Military quickly recognized the potential of fiber optics for improving its communications and tactical systems. In the early 1970's, the U.S. Navy installed a fiber optic telephone link aboard the U.S.S. Little Rock. The Air Force followed suit by developing its Airborne Light Optical Fiber Technology (ALOFT) program in 1976. Encouraged by the success of these applications, military R&D programs were funded to develop stronger fibers, tactical cables, ruggedized, high-performance components, and numerous demonstration systems ranging from aircraft to undersea applications.

Commercial applications quickly followed their military predecessors. In 1977, both AT&T and GTE installed fiber optic telephone systems in Chicago and Boston respectively. These successful applications led to the increase of fiber optic telephone networks. Originally designed around multimode graded-index fiber, by the early 1980's, single-mode fiber operating in the 1310 nm and later the 1550 nm wavelength windows became the standard fiber installed for these networks. For example, by 1983, British Telecom switched their entire phone system over to single-mode fiber. Initially, computers, information networks, and data communications were slower to embrace fiber, but today they too find application in a transmission system, using lighter weight cable, that resists lightning strikes and carries more information faster and over longer distances.

The broadcast industry also embraced fiber optic transmission. In 1980, broadcasters of the Winter Olympics, in Lake Placid, New York, requested a fiber optic video transmission system for backup video feeds. The fiber optic feed, because of its quality and reliability, soon became the primary video feed, making the 1980 Winter Olympics the first fiber optic television transmission. Later, at the 1994 Winter Olympics in Lillehammer, Norway, fiber optics transmitted the first ever digital video signal, an application that continues to evolve today.

In the mid-1980's the United States government deregulated telephone service, allowing small telephone companies to compete with the giant, AT&T. Companies like MCI and Sprint quickly went to work installing regional fiber optic telecommunications networks throughout the world. Taking advantage of railroad lines, gas pipes, and other natural rights of way, these companies laid miles fiber optic cable, allowing the deployment of these networks to continue throughout the 1980's. However, this created the need to expand fiber's transmission capabilities.

In 1990, Bell Labs transmitted a 2.5 Gb/s signal over 7,500 km without regeneration. His system used a soliton laser and an erbium-doped fiber amplifier (EDFA) that allowed the light wave to maintain its shape and density. In 1998, Bell Labs' success went one better as researchers transmitted 100 simultaneous optical signals, each at a data rate of 10 gigabits (giga means billion) per second for a distance of nearly 250 miles (400 km). In this experiment, dense wavelength-division multiplexing (DWDM) technology, which allows multiple wavelengths to be combined into one optical signal, was used to increase the total data rate on one fiber to one terabit per second (10^{12} bits per second).

THE TWENTY-FIRST CENTURY AND BEYOND

Today, DWDM technology continues to develop. As the demand for data bandwidth increases, driven by the phenomenal growth of the Internet, the move to optical networking is the focus of new technologies. At this writing, nearly half a billion people have Internet access and use it regularly. Some 40 million or more households are "wired." The world wide web already hosts over 2 billion web pages, and according to estimates people upload more than 3.5 million new web pages everyday.

Figure 1.4: Projected Internet Traffic Increases

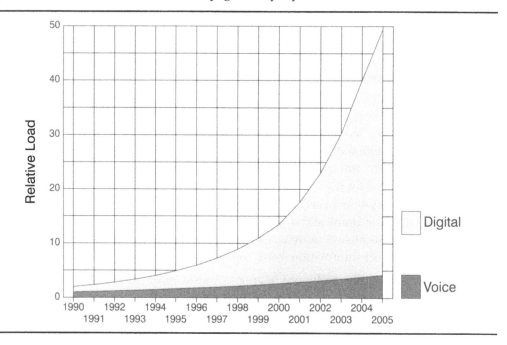

The important factor in these developments is the increase in fiber transmission capacity, which has grown by a factor of 200 in the last decade. Figure 1.5 illustrates this trend.

Figure 1.5: The Growth of Optical Fiber Transmission Capacity

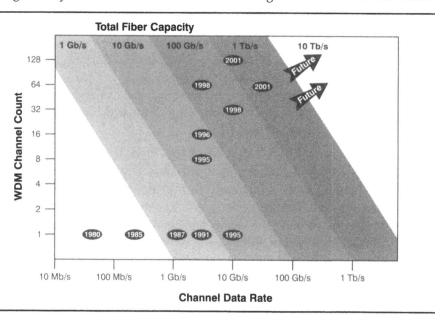

Because of fiber optic technology's immense potential bandwidth, 50 THz or greater, there are extraordinary possibilities for future fiber optic applications. Already, the push to bring broadband services, including data, audio, and especially video, into the home is well underway. Broadband service available to a mass market opens up a wide variety of interactive communications for both consumers and businesses, bringing to reality interactive video networks, interactive banking and shopping from the home, and interactive distance learning. Video-on-demand hit a brick wall in coax cable which excelled at carrying the same signal to everyone, but failed as a means to route switched signals intended for a specific generation. The "last mile" for optical fiber goes from the curb to the television set top, allowing video-on-demand to become a reality.

The next generation of fiber optic technology will probably involve a revamping of existing network architectures, to develop cost-effective "all-optical" networks. The move to all-optical networks would reduce system complexity and cost, eliminating the need to convert transmission signals between the electrical and optical domains. As Internet traffic continues to double every three to four months, the demand for fast, reliable connections will grow with it. Where fiber-to-the-curb seemed like science fiction ten years ago, fiber-to-the-desktop is a reality today. The optical layer of today's broadband networks seems remote from the end user, but wavelength-to-the-desktop, all-optical terminals connected directly to the end user's desktop computer, may not be as far away as once imagined.

Chapter Summary

- John Tyndall demonstrated in 1870 that light used internal reflection to follow a specific path.
- William Wheeling patented his method of light transfer, "piping light," in 1880.
- The photophone, the world's first optical AM audio link, was developed in 1880 by Alexander Graham Bell.
- The fiberscope, which used the first practical glass-coated fiber, was concurrently developed in the 1950's by Narinder Kapany and colleagues and Brian O'Brien.
- The term "fiber optics" was coined by Narinder Kapany in 1956.
- The laser gained acceptance in scientific circles during the late 1950's and early 1960's.
- In 1966, Charles Kao and Charles Hockam, at England's Standard Telecommunication Laboratory, proposed that optical fiber could carry laser signals.
- AT&T and GTE installed the first fiber optic telephone systems in 1977.
- The first fiber optic video transmission took place during the 1980 Winter Olympics in Lake Placid, New York.
- The first fiber optic digital video transmission took place during the 1994 Winter Olympics in Lillehammer, Norway.
- A Bell Lab's group broke the terabit per second barrier in 1998 when they successfully transmitted 100 simultaneous 10 Gb/s signals on one fiber over a distance of 400 km using DWDM technology.
- Broadband networks offering interactive communication services have become commonplace.
- All-optical networks may be the next generation solution to increasing bandwidth demands as a result of the growth of the Internet.

Selected References and Additional Reading

Das, Saswato and Kirk Saville. "Trends: Photonics." *Bell Labs Technology* 2.2 (1998): 1-5.

Designer's Guide to Fiber Optics. Harrisburg, PA: AMP, Incorporated, 1982.

Hecht, Jeff. *City of Light: The story of fiber optics.* Oxford, England: Oxford University Press, 1999.

—. *Understanding Fiber Optics.* 2nd edition. Indianapolis, IN: Sams Publishing, 1993.

Just the Facts. New Jersey: Corning Incorporated, 1992.

Palladino, John R. *Fiber Optics: Communicating By Light.* Piscataway, NJ: Bellcore, 1990.

FIBER OPTIC FUNDAMENTALS

THE NATURE OF LIGHT

In order to understand some of the more complex components used in modern, high-performance, fiber optic transmission systems, one should have a good understanding of the nature of light. Many of light's properties are not obvious in our everyday lives, yet these properties are vital to the success of many modern fiber optic components. An important characteristic of light is the fact that light acts like a wave. (It can also act like a particle, but that is not pertinent to this discussion.) Light travels as an oscillating wave of electric and magnetic fields. As a result, two phenomena affect the behavior of light: interference and polarization, both important fundamentals in understanding and designing modern fiber optic components.

Interference

Interference forms the basis of many modern fiber optic components, including fiber Bragg gratings, optical filters built directly into the fiber; lithium niobate modulators, used to modulate the laser or LED externally rather than by internal circuitry; and many types of bulk filters, devices used in wavelength-division multiplexing.

In Figure 2.1, the top two curves represent two light waves traveling along the same path. The figure illustrates only the electrical fields. The distance between the crests or troughs of the waves is the light's wavelength. This is analogous to color or frequency in later discussions. The two light waves shown are also "in phase." This means the peaks and the valleys of the two waveforms are perfectly aligned. When two light waves have the same wavelength and are in phase, they add together to produce a wave with the sum of the amplitudes of the input waves. This is known as constructive interference.

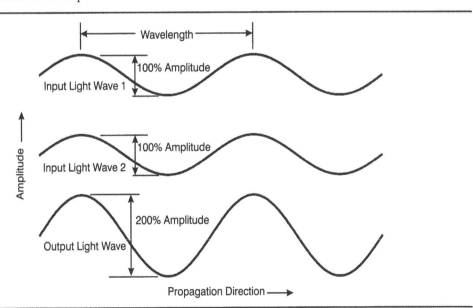

Figure 2.1:
Constructive
Interference

Figure 2.2 illustrates destructive interference, showing two light waves with the same wavelength but now "out of phase." The light waves shifted so that the peak on the first light wave aligns with the valley in the second light wave. In this case, the two input light waves almost cancel each other out, resulting in a very small output light wave.

Figure 2.2:
Destructive
Interference

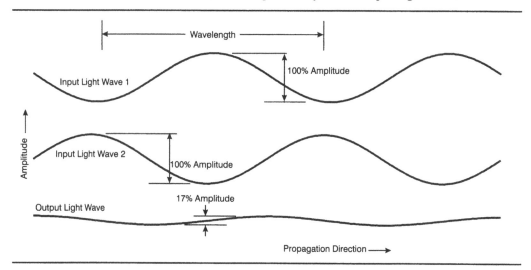

What causes interference in real systems? It usually occurs when a light wave combines with a delayed version of the same light wave. This principle governs lithium niobate modulators. Lithium niobate is an optically transparent crystal with the ability to change the delay of light in response to a change in an applied electric field. See Chapter 7 for a complete description of these modulators.

Polarization

Light waves travel along two planes separated by 90°. Figure 2.3 shows a polarized light wave traveling along the X axis. It consists of two components; the component labeled E represents the varying electric field and is known as the E-vector. The component labeled B represents the varying magnetic field and is known as the B-vector. This light wave shown is said to be vertically polarized because the E-vector is along the vertical axis. By convention, whichever axis the E-vector lies on is the plane of polarization. An unpolarized light wave would be a jumble of horizontally and vertically polarized light waves traveling together.

Figure 2.3: Light
Waves Showing E and
B Vectors

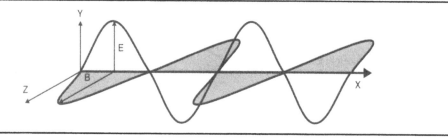

Polarized sunglasses most commonly make use of this principle. Their design allows the passage of only one polarization of light. Most reflected light has a horizontal polarization. The lens in the polarized sunglasses is vertically polarized, so it blocks the horizontally polarized reflection.

The Electromagnetic Spectrum

Light is organized into what is know as the electromagnetic spectrum. In this scheme, light refers to more than the portion of the electromagnetic spectrum that is visible to the human eye. The electromagnetic spectrum is composed of visible and near-infrared light like that transmitted by fiber, and all other wavelengths used to transmit signals such as AM and FM radio and television. Figure 2.4 illustrates the electromagnetic spectrum. As one can see, only a very small part of the spectrum is perceived by the human eye as light.

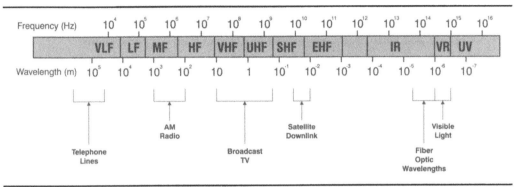

Figure 2.4: Electromagnetic Spectrum

Fiber optic wavelengths are measured in nanometers (the prefix nano meaning one-billionth) or microns (the prefix micro meaning one-millionth). Wavelengths for fiber optic applications can be broken into two main categories: near-infrared and visible. Visible light, as defined by the human eye, ranges in wavelengths from 400 to 700 nanometers (nm) and has very limited uses in fiber optic applications, due to the high optical loss. Near-infrared wavelengths range from 700 to 1,700 nanometers; these wavelengths are almost always used in modern fiber optic systems.

APPLYING LIGHT

Light signals have been used to communicate for as long as lighthouses have warned sailors away from treacherous shorelines. This application seems simplistic compared to today's possibilities. As countries continue to develop networks for global communication, fiber optics offers a method of transmission that allows for clearer, faster, more efficient communication than copper. A fiber optic communication system holds many advantages over a copper wire system. For example, while a simple two-strand wire can carry a low-speed signal over a long distance, it cannot send high-speed signals very far. Coaxial cables can better handle high-speed signals but still only over a relatively short distance. An advantage of copper cables, that fiber cannot duplicate, is the ability to transmit AC or DC power in addition to communication signals. Fiber optics holds a great advantage over copper media because it can handle high-speed signals over extended distances. Other advantages of fiber include:

Immunity from Electromagnetic (EM) Radiation and Lightning: Because the fiber itself is made from dielectric (nonconducting) materials, it is unaffected by EM radiation. The electronics required at the end of each fiber, however, are still susceptible and require shielding. Immunity from EM radiation and lightning was initially most important to the military. In terms of secure communications, fiber itself is not inherently secure, but it does not normally emit any EM radiation that can be readily detected. Immunity to EM radiation is important in modern aircraft designs that have composite skins; these nonconducting skins do not shield the electronics or wiring from EM fields or radiation. Lightning immunity is a key reason to use fiber optic devices in commercial security and intelligent transportation systems, because these systems are usually dispersed over a wide area making them susceptible to damage from lightning strikes and interference. Immunity from EM radiation is an important factor in choosing fiber to upgrade existing communication systems. The fiber can often be run in the same conduits that currently carry power lines, simplifying installation.

Lighter Weight: This feature refers to the optical fiber itself. In real world applications, copper cables can often be replaced by fiber optic cables that weigh at least ten times less. For long distances, a complete fiber optic system (optical fiber and cable, plus the supporting electronics) also has a significant weight advantage over copper systems. This is often not true for short systems, however, because fiber optic systems almost always require more elaborate, and thus larger and heavier electronics than copper systems.

Higher Bandwidth: Fiber has higher bandwidth than any alternative available. The CATV industry in the past required amplifiers every thousand feet or so on their supertrunks when copper cable was used. This is due to the limited bandwidth of the copper cable. A modern fiber optic system can carry the same signals with similar or superior signal quality for 50 miles or more without needing intermediate amplification (repeaters). Even at that, most modern fiber optic communication systems use less than a few percent of fiber's inherent bandwidth.

Better Signal Quality: Because fiber is immune to EM interference, has lower loss per unit distance, and wider bandwidth, signal quality is usually substantially better compared to copper.

Lower Cost: This has to be qualified. Fiber certainly costs less for long distance applications. However, for signal transmission requirements over distances of a few feet, copper is cheaper and probably always will be. The cost of fiber itself is cheaper per unit distance than copper if bandwidth and transmission distance requirements are high; however, the cost of the electronics and electro-optics at the end of the fiber can be substantial. The price of copper is such today that for very long distance links, converting to a fiber optic system can often be completely paid for by the salvage value of the removed copper. Fiber and copper systems can be compared by finding a break-even distance. At distances shorter than the break-even distance, copper is cheaper and vice versa. In the mid 1980's the break-even distance was 10 km or more. Today, the break-even distance is often less than 100 meters. Most of this gain has come from the reduced cost of fiber optic systems and components. Some of the gain is also due to the fact that the cost of copper has increased over the same period of time.

Easily Upgraded: The limitation of fiber optic systems today, and for many years to come, is the electronics and electro-optics used on each end of the fiber. The fiber itself usually has much more transmission capability, especially higher bandwidth, than is being utilized. Once fiber is installed, particularly a single-mode fiber, advances in electronics and electro-optics can be readily incorporated using the installed fiber.

Ease of Installation: Many newcomers to fiber optics are often concerned about glass being very brittle and prone to breakage. In fact, glass is many times stronger than steel, and optical fibers are so small that they are very flexible. A good quality fiber optic cable incorporates strain relief materials as well as bend limiters that make the cable very hardy. Copper coax, is in fact, much more fragile than a fiber optic cable. Copper coax cables are prone to kinks and deformities that will permanently degrade the performance of the cable. Glass however, will not deform or kink. Thus, it can usually take much more abuse than copper.

TYPICAL FIBER OPTIC COMMUNICATIONS SYSTEM

Modern fiber optic transmission systems can be extraordinarily complex as the data rates, channel counts, and transmission distances increase. However, the basic principles behind fiber optic transmission are relatively simple. As shown in Figure 2.5, fiber optic links contain three basic elements: the transmitter (Tx) that allows for data input and outputs an optical signal, the optical fiber that carries the data, and the receiver (Rx) that decodes the optical signal to output the data.

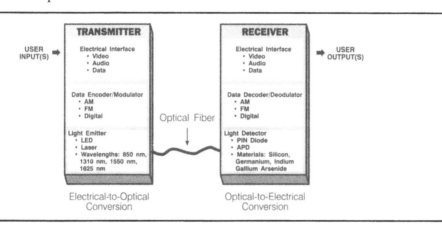

Figure 2.5: Elements of a Fiber Optic Link

The transmitter in Figure 2.5 uses an electrical interface, either video, audio, data, or other forms of input, to encode the user's information through modulation. Three forms of modulation are typically used: amplitude modulation (AM), frequency modulation (FM), and digital modulation. The electrical output of the modulator is usually transformed into light either by means of a light-emitting diode (LED) or a laser diode (LD). The wavelengths of these light sources range from 780 nm to 1625 nm for most fiber optic applications.

The receiver in Figure 2.5 decodes the light signal back into electrical signals. Two types of light detectors are typically used: the PIN photodiode or the avalanche photodiode (APD). Typically, these detectors are made from silicon (Si), indium gallium arsenide (InGaAs), or germanium (Ge). The detected and amplified electrical signal is then sent through a data decoder or demodulator that converts the electrical signals back into video, audio, data, or other forms of user input.

The next level of complexity for fiber optic transmission systems involved the addition of multiple wavelengths of light on each fiber. Each wavelength of light is a distinct "color" that can be combined for transmission and subsequently separated out at the receiver end by using various optical filters. This technique is referred to as wavelength-division multiplexing (WDM). Wavelength is usually denoted by the Greek symbol, lambda or λ, in fiber optic systems. Often the λ symbols have subscripted numbers denoting different wavelengths, e.g., λ_1. Figure 2.6 shows a basic WDM system transmitting two wavelengths in the same direction on the fiber.

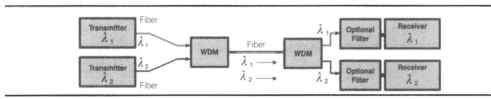

Figure 2.6: Basic WDM System

As the load from Internet traffic increased, WDM systems evolved into systems that could carry a multitude of wavelengths (colors). These systems were called dense wavelength-division multiplexing (DWDM) systems. Erbium-doped fiber amplifiers (EDFA) made DWDM systems practical. The EDFA allowed the multitude of wavelengths on the fiber to be boosted simultaneously, making the systems very economical. Early systems carried four to eight wavelengths and then later, sixteen. The increase in capacity per fiber was dramatic but still insufficient to handle the exponentially growing Internet traffic load. Figure 2.7 shows a simple four wavelength DWDM fiber optic transmission system. An EDFA is included to increase the possible transmission distance.

Figure 2.7: Basic Four Channel DWDM System

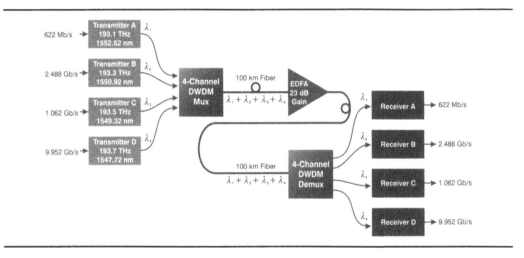

Figure 2.8 shows many key elements associated with a modern fiber optic transmission system such as external modulators, forward error correction, solitons, and Raman amplifiers. These more advanced concepts will be discussed fully in chapters 7, 9, and 14 of this book.

Figure 2.8: Modern DWDM Long-haul Communication System

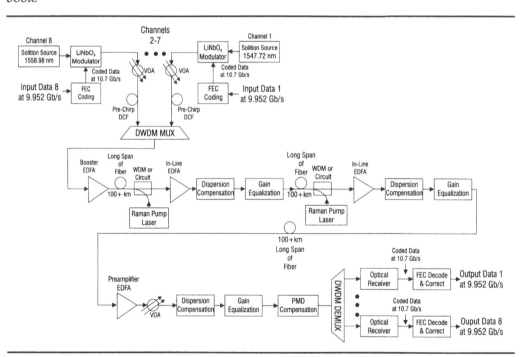

In any transmission system, fiber provides a private pipeline that can carry huge amounts of data. Alternatives to fiber optics include over-the-air broadcast and hard-wired copper wires carrying electrons. Figure 2.9 explains these three schemes for transmitting information from one point to another.

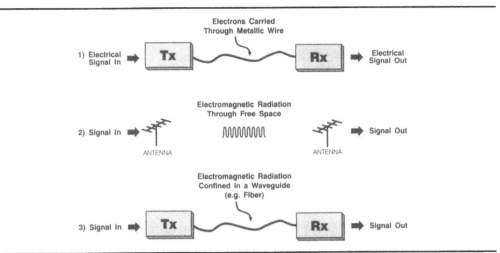

Figure 2.9:
Transmission Schemes

Metallic transmission, the first scheme, uses a copper wire or coaxial cable to carry a modulated electrical signal containing information. This method allows for a limitless number of private channels (assuming you have that much copper cable), but each channel has limited information and distance capability due to the inherent characteristics of copper cable. The second scheme for moving information between two locations is free-space transmission. This is how radio signals and over-the-air TV signals are received. Free-space transmission has the advantage of providing very large bandwidth capability as well as long distance capability, but it does not provide a private channel. Also, the free-space spectrum is a finite, limited, and costly commodity. It cannot provide the millions of high-speed communication channels required by the global information age. The last scheme is waveguide transmission. This describes optical fiber transmission. A waveguide (optical fiber) confines the electromagnetic radiation (light) and moves it along a prescribed path. Optical fiber offers the best of both metallic and free-space transmission. It has the key advantage of metallic transmission, the ability to carry a signal from point A to point B without cluttering the limited free-space electromagnetic spectrum; however, fiber does not have the disadvantage of metallic transmission: very limited bandwidth and data rate.

FIBER OPTIC COMPONENTS

Optical Fiber: Optical fibers are extremely thin strands of ultra-pure glass designed to transmit light signals from a transmitter to a receiver. These signals represent electrical signals that include, in any combination, video, audio, or data information. Figure 2.10 shows the general cross-section of an optical fiber. The fiber consists of three main regions. The center of the fiber is the core. This part of the fiber actually carries the light. It ranges in diameter from 9 microns (μm) to 100 microns in the most commonly used fibers. The next region, the cladding, surrounds the core and confines the light in the core. The cladding typically has a diameter of 125 microns. A key design feature of all optical fibers is that the refractive index of the core is higher than the refractive index of the cladding. Both the core and cladding are usually doped glass materials. The outer region of the optical fiber is called the coating or buffer. The buffer, typically a plastic material, provides protection and preserves the strength of the glass fiber. Usual diameters for the buffer are 250 microns, 500 microns, and 900 microns. Optical fiber is discussed extensively in Chapter 3.

Figure 2.10: Cross-section of an Optical Fiber

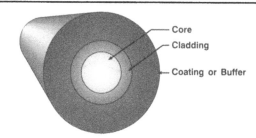

Core

Cladding

Coating or Buffer

Light Emitters: The two types of light sources used in fiber optics are light-emitting diodes (LED's) and laser diodes (LD's). LED's may be surface-emitting or edge-emitting sources. The surface-emitting LED (SLED) emits light over a very wide angle. This type of light source is often called a Lambertian emitter because of the nature of the emission pattern. This broad emission angle is attractive for use as an indicating LED because of the wide viewing angle, but is a detriment for fiber optic uses. Because the emitting angle is so large, it is difficult to focus more than a small amount of the total light output into the fiber core. The key advantage of surface-emitting LED's is their low cost, making these light emitters the dominant type in use. The second type of light emitter is the edge-emitting LED (ELED). This LED type has a much narrower angle of light emission and also has a smaller emitting area. This allows a larger percentage of the total light output to be focused into the fiber core. ELED's are also generally faster than their surface-emitting cousins. The disadvantage of ELED's is that they are very temperature sensitive compared to SLED's. The last light emitter type is the laser diode. A laser has a very narrow emission angle, and the emitting spot is very small, usually only a few microns in diameter. Because of the small emission angle and emitting spot, a very high percentage, often more than 50%, of the output light can be focused into the fiber core. The laser diode is the fastest of these three emitter types. Figure 2.11 illustrates the emission patterns of these sources.

Figure 2.11: Sources and Emission Patterns

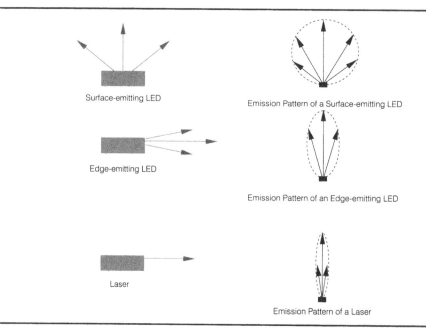

Surface-emitting LED

Emission Pattern of a Surface-emitting LED

Edge-emitting LED

Emission Pattern of an Edge-emitting LED

Laser

Emission Pattern of a Laser

Many characteristics are considered when making the decision to use LED's or LD's in fiber optic systems. Some basic considerations include center wavelength, spectral width (the full range of wavelengths around the center wavelength, often called the FWHM, Full Width Half Maximum), and optical output power of the light source. The spectral width is important as larger values bring with them an increased possibility of dispersion problems, limiting bandwidth. The optical output power of the light source is also important in that it must be neither too weak nor too strong. A weak source will not provide enough power to transmit a light signal through a usable length of optical fiber, while a source that is too strong could cause distortion of the signal by overloading the receiver. These and other light source characteristics will be explored at length in Chapter 5.

Detectors: Detectors convert optical power into an electrical signal. These may be positive-intrinsic-negative (PIN) photodiodes or avalanche photodiodes (APD's). The basic difference between the two is that the APD provides a good deal of internal gain, simplifying the task of signal amplification. As is almost always the case with engineering trade-offs, with an advantage comes a disadvantage. APD's are more expensive and require extensive additional support electronics because of the need to provide a high-voltage bias supply that must be temperature compensated. PIN diodes, on the other hand, do not require this additional support circuitry and are very economical, but they require more complex amplifier stages. PIN detectors are by far the most economical option, which gives them the largest share of the market. APD's are becoming more important as data rates and span lengths increase. Light detectors are discussed in detail in Chapter 6.

Interconnection Devices: An interconnection device is any component or technique used to connect a fiber or a fiber optic component to another component or fiber. Interconnection devices provide both light junctions and mechanical junctions for interconnecting fiber optic systems. As light junctions, they provide outputs or inputs for signal sources. Mechanically, they hold the connections in place. Interconnection devices include connectors and splices. Some uses for interconnection devices are:

- Interfaces between local area networks and devices
- Patch panels
- Portable military communication links
- Network-to-terminal connections
- Connections between recording equipment, cameras, and sound equipment in portable studios

Interconnection devices are treated in detail in Chapter 8.

In addition to these fiber optic components, this guide will discuss active devices for optical amplification (Chapter 7), passive devices for WDM transmission and switches (Chapter 9), system design (Chapter 10), applications for optical fiber (Chapter 11), video transmission (Chapter 12), data transmission (Chapter 13), the theoretical and practical limitations of optical fiber (Chapter 14), testing and measurement techniques (Chapter 15), and the future of the fiber optics industry (Chapter 16).

SOME USEFUL TERMINOLOGY

A few more terms should be explained before this book proceeds any further, especially as the discussion on digital transmission evolves.

Bit: A bit is a single digit in a binary number system, either a zero or a one. (The base ten counting system that we use in our everyday lives is so called because each digit can be any value from zero to nine.) The binary number system and the bit are the basis of all modern computers, calculators and communication systems. The bit can be thought of as a type of Morse code where any symbol (e.g., a letter or number) may be represented by a series of dots and dashes, corresponding to a zero and a one in binary.

Data Rate: Data rate describes the rate at which symbols (letters or numbers), are used to transmit a message. A modern fiber optic system can easily transmit at a data rate of 10 Gbit/s. This means that 10 billion bits (zeroes or ones) move from point A to point B every second.

Digital Transmission: The process of moving bits from point A to point B is referred to as digital transmission.

Chapter Summary

- Because light acts like a wave in many instances, effects such as interference and polarization can impact modern fiber optic systems.
- Constructive interference adds to the light wave output, but destructive interference diminishes the light wave output.
- The electromagnetic spectrum includes all the wavelengths or colors of light, not just colors visible to the human eye.
- Fiber optic systems holds many advantages over conventional copper wire and coax cable systems, including EMI immunity, lighter weight, higher bandwidth, lower cost, and better signal quality.
- Fiber optic components transmit information by turning electronic signals into light and light signals back into electronic signals.
- Wavelengths of light used in fiber optic applications include near-infrared and visible.
- The three basic elements of a fiber optic link are the transmitter, the receiver, and the optical fiber.
- The most common transmission schemes are metallic transmission, free-space transmission, and waveguide transmission.
- Optical fibers are extremely thin strands of ultra-pure glass designed to transmit light signals from a transmitter to a receiver.
- The two types of light sources used in fiber optics are light-emitting diodes (LED's) and laser diodes (LD's).
- Two types of detectors associated with fiber optics are the PIN photodiode, and the avalanche photodiode (APD).
- An interconnection device is any component or technique used to connect a fiber or fiber optic component to another component or another fiber. These devices include connectors, splices, couplers, splitters, switches and wavelength-division multiplexers (WDM's).

Selected References and Additional Reading

Baack, Clemens. *Optical Wideband Transmission Systems.* Florida: CRC Press, Inc., 1986.

Hecht, Jeff. *Understanding Fiber Optics.* 2nd edition. Indianapolis, IN: Sams Publishing, 1993.

Sterling, Donald J. *Amp Technician's Guide to Fiber Optics*, 2nd Edition. New York: Delmar Publishers, 1993.

OPTICAL FIBER

By 1950, the challenge to scientists studying optical fiber transmission was not whether light could carry information, but whether a glass conduit could be developed that was pure enough to keep losses below 20 dB/km. A flexible glass-coated glass fiber served as a suitable transmission medium for the fiberscope, but losses remained unworkably high for communication applications. Scientists persevered.

In 1970, Corning scientists Robert Maurer, Donald Keck, and Peter Schultz developed a fiber with a measured attenuation of less than 20 dB/km. It was the purest glass ever made, and the breakthrough led to the commercialization of fiber optics for communication applications. Corning's success was the result of a new process for manufacturing optical fiber which they called inside vapor deposition (IVD). Instead of melting the raw silica, the way most glass is made, they formed the glass from vaporized chemicals which were deposited inside a silica tube. The outer tube became the cladding, and the core of the fiber formed within. An inherent disadvantage to this method was that the cladding was formed using the traditional method of making glass. Thus impurities still existed in the cladding and reduced the purity of the overall fiber. Back to the drawing board they went, and before long, Corning scientists had developed a method of outside vapor deposition (OVD) which formed the entire fiber from ultra-pure, vapor deposited chemicals. Today scientists still experiment with ever purer forms of optical fiber, and losses as low as 0.2 dB/km at 1550 nm are not uncommon.

FIBER ERA's

In the early days of fiber optics, optical fiber was simple. Everyone assumed that fiber had infinite bandwidth and would meet mankind's communication needs into the foreseeable future. As the need arose to send information over longer distances, the fiber community developed additional wavelength windows that would allow longer and longer transmission. The third window, 1550 nm, seemed to be the ultimate answer. With losses of only 0.2 dB/km, it appeared adequate for any imaginable application. Millions of kilometers of fiber were installed around the world creating a high-speed communication network that would surely last for years.

But then a few years ago, the Internet happened, and simultaneously, demand for many video channels exploded. The old days of sending 140 Mb/s over a 50 km fiber length ended. More bandwidth was required, and fast! Some researchers felt that terabit data rates were possible over fiber. Theory confirmed that conjecture, but could conventional electronics keep up? So far the answer has been an emphatic NO. It appears that electronics will handle up to 40 Gb/s data rates, but not much more. But fiber could support additional types of multiplexing beyond purely electronic schemes such as time-division multiplexing (TDM). It could also support transmission of many different colors or wavelengths of light.

In the last few years, researchers have developed affordable dense wavelength-division multiplexing (DWDM) technology. Suddenly, ten, 20, or even 80 or more simultaneous data streams can combine onto a single fiber, each transmitted at a slightly different color of light. The impact of DWDM on the telephony, video, and data communications industries will be staggering. Now currently installed fibers can carry up to two orders of magnitude more data than they could even a few years ago.

MANUFACTURE OF OPTICAL FIBER

Modified chemical vapor deposition (MCVD), another term for the IVD method, and outside vapor deposition (OVD) are the two predominant methods for manufacturing optical fiber. The MCVD process involves depositing ultra-fine, vaporized raw materials into a pre-made silica tube. The soot that develops from this deposition is consolidated by heating. The resulting preform or blank is heated and drawn into a hair-thin optical fiber. Figure 3.1 illustrates the MCVD process.

Figure 3.1:
Manufacturing Optical Fiber (MCVD Process)

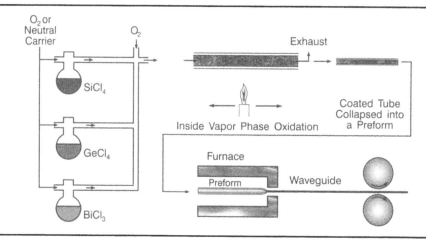

The optical fiber is encased in several protective layers to ensure integrity under various conditions. The first layer is applied to the glass fiber as it is drawn from the preform. This coating is generally made of ultraviolet-curable acrylate or silicone, and it serves as a moisture shield and as mechanical protection during the early stages of cable production. A secondary buffer may be extruded over the primary coating to improve strength.

The OVD process involves deposition of raw materials onto a rotating rod. This occurs in three steps: laydown, consolidation, and draw. During the laydown step, a soot preform is made from ultra-pure vapors of silicon tetrachloride and germanium tetrachloride. The vapors move through a traversing burner and react in the flame to form soot particles of silica and germanium oxide. These particles are deposited on the surface of the rotating target rod. When the deposition is complete, the rod is removed, and the deposited material is placed into a consolidation furnace. The water vapor is removed, and the preform is collapsed to become dense, transparent glass. This preform is drawn into a continuous strand of glass fiber in much the same way that taffy is stretched and thinned. By varying the mixture of gases throughout the process, the preform has a step-index or a graded-index of refraction. Figure 3.2 displays the method of drawing optical fiber.

Figure 3.2: Optical Fiber Draw Process

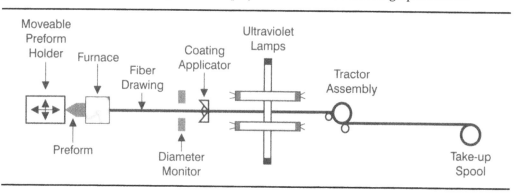

PRINCIPLES OF OPERATION

The principle of total internal reflection governs the operation of optical fiber. We don't normally encounter this process in our everyday lives. When we look in a mirror each morning, the image we see represents only 90% of the light reflecting back at us. Total internal reflection reflects 100% of the light.

One can observe this principle when viewing a fish tank. Figure 3.3 shows the top view of a fish tank complete with an occupant and an observer. From a viewpoint behind the position of the fish, both the observer and the fish can see light bulb "A." However, light bulb "B" cannot be seen because of the total internal reflection that occurs. In this example, the water has a refractive index of 1.33, and the air has a refractive index of about 1.00. (Actually there is a glass wall between the water and air with a refractive index of 1.45, but this doesn't materially affect the experiment.)

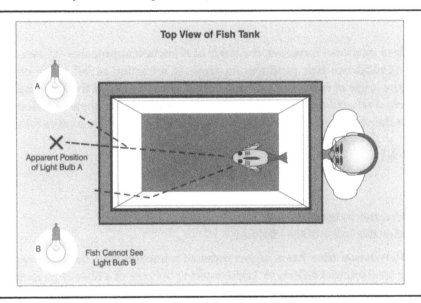

Figure 3.3: Principle of Total Internal Reflection

Figure 3.4 demonstrates the equations involved in this principle.

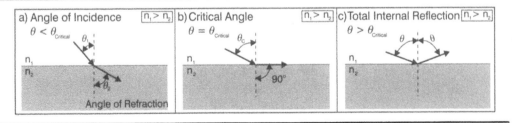

Figure 3.4: Snell's Law—Equations for Total Internal Reflection

In Figure 3.4, the upper, lighter region of each frame always has a higher refractive index than the lower, darker region. The refractive index of the upper region is designated n_1 while the lower region refractive index is n_2. Figure 3.4a shows a case where the angle of incidence is less than the critical angle. Note that the angle the light travels changes at the interface between the higher refractive index, n_1 region, and the lower refractive index, n_2 region. In Figure 3.4b, the angle of incidence has increased to the critical angle. At this angle the light ray travels parallel to the interface region. In Figure 3.4c, the incidence angle has increased to a value greater than the critical angle. In this case 100% of the light reflects at the interface region.

Total internal reflection occurs because light travels at different speeds in different materials. A dimensionless number called the index of refraction or refractive index, characterizes the different mediums through which the light is traveling. The index of refraction is the ratio of the velocity of light in a vacuum (c) to its velocity in a specific medium (v).

Eq. 3.1
$$n = \frac{c}{v}$$

As light passes from one medium to a medium with a different index of refraction, it is bent, or refracted. If light passes from a medium with a lower index of refraction to one with a higher index of refraction the light is bent toward the normal, and if the light passes from a higher to lower index of refraction the light is bent away from the normal. Snell's Law, in Figure 3.4, determines the amount the light is bent and is given by:

Eq. 3.2
$$n_1 \sin\Theta_1 = n_2 \sin\Theta_2$$

As the angle of incidence increases, the angle of refraction approaches 90° (see Figure 3.4). The angle of incidence that produces an angle of refraction of 90° is the critical angle. Increasing the angle of incidence past the critical angle results in total internal reflection. In total internal reflection the angle of incidence is equal to the angle of reflection. This is the basis for the operation of optical fiber. The critical angle is calculated as follows:

Eq. 3.3
$$\Theta_c = \sin^{-1}\left(\frac{n_2}{n_1}\right)$$

Where:

   n_1 = Refractive index of the core.
   n_2 = Refractive index of the cladding.

The core of an optical fiber has a higher index of refraction than the cladding ($n_1 > n_2$), allowing for total internal reflection. Light entering the core of a fiber at an angle sufficient for total internal reflection travels down the core reflecting off the interface between the core and the cladding. Light entering at an angle less than the critical angle is refracted into the cladding and lost.

There is an imaginary cone of acceptance with an angle, α, determined by the critical angle. This is related to a parameter called the numerical aperture (NA) of the fiber. NA describes the light gathering capability of fiber and is given as:

$$NA = \sin\alpha = \sqrt{n_1^2 - n_2^2}$$
Eq. 3.4
$$\text{and}$$
$$\alpha = \sin^{-1}\left(\sqrt{n_1^2 - n_2^2}\right)$$

Figure 3.5 illustrates numerical aperture and shows the location of angle α.

Figure 3.5: Numerical Aperture

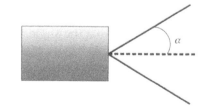

Here is an example of the above parameters for a real fiber. Let us assume that the core refractive index is 1.47, and the cladding refractive index is 1.45.

So:

$n_1 = 1.47$
$n_2 = 1.45$

From Equation 3.1 we can find the speed at which light travels in the fiber.

Eq. 3.5 $\qquad v = \dfrac{c}{n} = \dfrac{(2.998 \cdot 10^8 \text{ m/s})}{1.47} = 2.039 \cdot 10^8 \text{ m/s}$

This can be inverted to find that it takes light 4.903 ns to travel one meter in an optical fiber.

From Equation 3.3 we can calculate the critical angle, Θ_c:

Eq. 3.6 $\qquad \Theta_c = \sin^{-1}\left(\dfrac{1.45}{1.47}\right) = 80.5°$

From Equation 3.4 we can calculate the numerical aperture to be:

Eq. 3.7 $\qquad NA = \sqrt{(1.47^2 - 1.45^2)} = 0.2417$

Finally, also from Equation 3.4 we can calculate α.

Eq. 3.8 $\qquad \alpha = \sin^{-1}(0.2417) = 13.98°$

Table 3.1 lists the refractive indices and propagation times of a variety of mediums.

Medium	Refractive Index	Propagation Time
Vacuum	1.000	3.336 ns/m
Air	1.003	3.346 ns/m
Water	1.333	4.446 ns/m
Fused Silica	1.458	4.863 ns/m
Copper Coax Cable (RG-59/U)	N/A	5.051 ns/m

Table 3.1: Refractive Indices and Propagation Times

While the propagation times of optical fiber and copper coax cable appear almost identical, they are determined by different factors. In metallic cables, propagation delays are dependent on the cable dimensions and the frequency. In optical fiber, propagation delays are related to the refractive index of the material. Most applications are insensitive to the absolute propagation time delay caused by fiber. However, system designs that involve sending many synchronous digital signals over separate fibers require that all signals arrive at about the same time. Another application concerned with propagation time involves sending multiple signals over the same fiber using different wavelengths of light. Propagation time through a fiber is calculated as follows:

Eq. 3.9 $\qquad t = \dfrac{L \cdot n}{c}$

Where:

t = Propagation time in seconds.
L = Fiber length in meters.
n = Refractive index of the fiber core (approximately 1.47).
c = Speed of light (2.998×10^8 meters/second).

In doing a detailed analysis one must consider that the fiber length and refractive index are slightly temperature dependent, and the refractive index of the fiber, as shown in Figure 3.6, is also dependent on wavelength. It is difficult to obtain exact data for these variables, so one should consult the fiber manufacturer.

Figure 3.6: Refractive Index of Pure Fused Silica

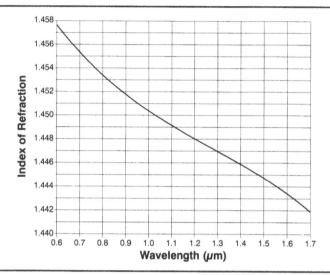

Now that we have discussed the principles of operation, it is time to describe how these principles operate in different types of optical fiber. As mentioned, there are two basic types of optical fiber: multimode fiber and single-mode fiber. Both are described in greater detail below.

MULTIMODE FIBER

Multimode fiber was first to be manufactured and commercialized. The term multimode simply refers to the fact that numerous modes or light rays are carried simultaneously through the waveguide. Modes result from the fact that light will only propagate in the fiber core at discrete angles within the cone of acceptance. This fiber type has a much larger core diameter, compared to single-mode fiber, allowing for the larger number of modes, and multimode fiber is easier to couple than single-mode optical fiber. Multimode fiber may be categorized as step-index or graded-index fiber.

Multimode Step-index Fiber

Figure 3.7 shows how the principle of total internal reflection applies to multimode step-index fiber. Because the core's index of refraction is higher than the cladding's index of refraction, the light that enters at less than the critical angle is guided along the fiber.

Figure 3.7: Total Internal Reflection in Multimode Step-index Fiber

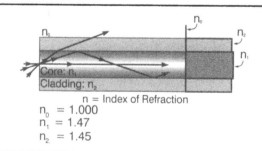

Three different rays of light are pictured traveling down the fiber. One mode travels straight down the center of the core. A second mode travels at a steep angle and bounces back and forth by total internal reflection. The third mode exceeds the critical angle and is refracted into the cladding and lost as it escapes into the air. Intuitively, it can be seen that the second mode travels a longer distance than the first mode, causing the two modes to arrive at separate times. This disparity between arrival times of the different light rays is known as dispersion, and the result is a muddied signal at the receiving end. Dispersion will be treated in detail later in this chapter; however, it is important to note that high dispersion is an unavoidable characteristic of multimode step-index fiber.

Multimode Graded-index Fiber

To compensate for the dispersion inherent in multimode step-index fiber, multimode graded-index fiber was developed. Graded-index refers to the fact that the refractive index of the core gradually decreases farther from the center of the core. The increased refraction in the center of the core slows the speed of some light rays, allowing all the light rays to reach the receiving end at approximately the same time, reducing dispersion.

Figure 3.8 shows the principle of multimode graded-index fiber. The core's central refractive index, n_A, is greater than that of the outer core's refractive index, n_B. As discussed earlier, the core's refractive index is parabolic, being higher at the center. As Figure 3.8 shows, the light rays no longer follow straight lines; they follow a serpentine path being gradually bent back toward the center by the continuously declining refractive index. This reduces the arrival time disparity because all modes arrive at about the same time. The modes traveling in a straight line are in a higher refractive index, so they travel slower than the serpentine modes. These travel farther but move faster in the lower refractive index of the outer core region.

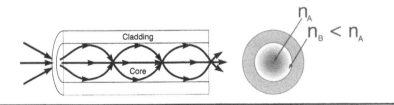

Figure 3.8: Multimode Graded-index Fiber

SINGLE-MODE FIBER

Single-mode fiber allows for a higher capacity to transmit information because they are able to retain the fidelity of each light pulse over longer distances, and they exhibit no dispersion caused by multiple modes. Single-mode fiber also enjoys lower fiber attenuation than multimode fiber. (Attenuation will be addressed further in the next section of this chapter.) Thus, more information can be transmitted per unit of time. Like multimode fiber, early single-mode fiber was generally characterized as step-index fiber meaning the refractive index of the fiber core is a step above that of the cladding rather than graduated as it is in graded-index fiber. Modern single-mode fibers have evolved into more complex designs such as matched clad, depressed clad and other exotic structures.

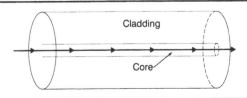

Figure 3.9: Single-mode Fiber

Single-mode fiber has disadvantages. The smaller core diameter makes coupling light into the core more difficult. The tolerances for single-mode connectors and splices are also much more demanding.

Single-mode Fiber Types in the 1990's

Single-mode fiber has gone through a continuing evolution for several decades now. As a result, there are three basic classes of single-mode fiber used in modern telecommunications systems. The oldest and most widely deployed type is non dispersion-shifted fiber (NDSF). These fibers were initially intended for use near 1310 nm.

Later, 1550 nm systems made NDSF fiber undesirable due to its very high dispersion at the 1550 nm wavelength. To address this shortcoming, fiber manufacturers developed, dispersion-shifted fiber (DSF), that moved the zero-dispersion point to the 1550 nm region. Years later, scientists would discover that while DSF worked extremely well with a single 1550 nm wavelength, it exhibits serious nonlinearities when multiple, closely-spaced wavelengths in the 1550 nm were transmitted in DWDM systems. Recently, to address the problem of nonlinearities, a new class of fibers were introduced. These are classified as non zero-dispersion-shifted fibers (NZ-DSF). The fiber is available in both positive and negative dispersion varieties and is rapidly becoming the fiber of choice in new fiber deployment. Figure 3.10 shows the dispersion of the common types of single-mode fibers. Summaries of each type follow.

Figure 3.10:
Dispersion Behavior of
Single-mode Fiber
Types

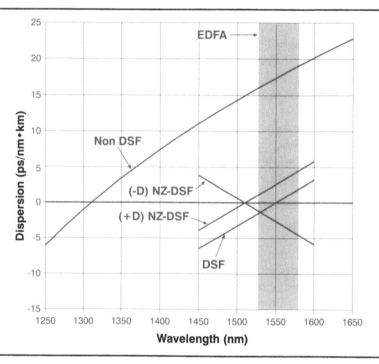

NDSF: Commonly referred to as standard single-mode silica fiber, this optical fiber is also known as non dispersion-shifted fiber (NDSF). SMF-28, made by Corning, is among the most popular NDSF deployed today. NDSF has an operating wavelength for zero chromatic dispersion (called λ_0) of 1310 nm. Chromatic dispersion causes a pulse to spread out as it travels along a fiber due to the fact that the different wavelength components that constitute the pulse travel at slightly different speeds in the fiber. The further away the wavelength is from λ_0 the greater the degree of dispersion and hence distortion. The fiber zero dispersion wavelength, or λ_0, is the wavelength at which chromatic dispersion is zero. Transmission wavelengths used with erbium-doped fiber amplifier systems (1550 nm window) undergo significant chromatic dispersion with NDSF and require dispersion compensation, particularly at 10 Gbit/s or higher data rates. Typical optical losses range from 0.21 to 0.25 dB/km.

DSF: Introduced in the early 1980's, dispersion-shifted fiber (DSF) minimized chromatic dispersion at 1550 nm. By changing the index profile and reducing the core radius, fiber designers were able to move λ_0 from the 1310 nm window to the 1540 to 1560 nm window. Standard DSF fiber has a λ_0 of 1557 $\pm$12.5 nm (i.e., 1544.5 nm to 1569.5 nm). Though effective in reducing chromatic dispersion effects, the positioning of λ_0 in close proximity to the operating wavelengths resulted in a tendency to encounter a nonlinear distortion effect called four wave mixing (FWM). This is especially troublesome in DWDM applications with more than eight wavelengths. Typical losses range from 0.25 to 0.30 dB/km.

NZ-DSF: NZ-DSF was developed to counteract the FWM limitations of DSF. The idea is to move λ_0 to either end of the 1550 nm band, thus ensuring that all of the wavelength channels have slightly different optical speeds in the fiber. Common types are TrueWave Classic (λ_0 < 1530 nm), TrueWave Plus (λ_0 = 1497 nm), TrueWave RS (λ_0 < 1452 nm) by Lucent, and SMF-LS (λ_0 > 1560 nm) by Corning. These fibers have a lower amount of dispersion as well as a higher tolerance to nonlinear distortion effects. This fiber is available with positive dispersion (+D) NZ-DSF and negative dispersion (-D) NZ-DSF. Figure 3.11 shows the dispersion along a run of fiber that alternates 20 kilometer lengths of (+D) NZ-DSF and (-D) NZ-DSF types.

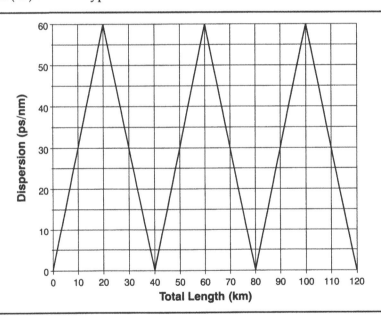

Figure 3.11: Dispersion for Alternating 20 km Lengths of (+D) NZ-DSF and (-D) NZ-DSF Fiber

Polarization-maintaining Fiber

One additional important variety of single-mode fiber is polarization-maintaining (PM) fiber. All other single-mode fibers discussed so far have been capable of carrying randomly polarized light. PM fiber is designed to propagate only one polarization of the input light. This is important for components such as external modulators that require a polarized light input. Figure 3.12 shows the cross-section of a type of PM fiber. This fiber contains a feature not seen in other fiber types. Besides the core, there are two additional circles called stress rods. As their name implies, these stress rods create stress in the core of the fiber such that the transmission of only one polarization plane of light is favored.

When PM fiber is connectorized, it is important that the stress rods line up with the connector, usually in line with the connector key. PM fiber also requires a great deal of care when it is spliced. Not only does the X, Y and Z alignment have to be perfect when the fiber is melted together, the rotational alignment must also be perfect so that the stress rods align exactly.

Figure 3.12: Cross-section of Polarization-maintaining Fiber

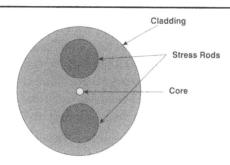

ATTENUATION

Attenuation (loss) is a logarithmic relationship between the optical output power and the optical input power in a fiber optic system. It is a measure of the decay of signal strength, or loss of light power, that occurs as light pulses propagate through the length of the fiber. The decay along the fiber is exponential and can be expressed as:

Eq. 3.10
$$P(L) = P_0 \cdot e^{-\alpha' L}$$

Where:

P(L) = Optical power at distance L from the input.
P_0 = Optical power at fiber input.
α' = Fiber attenuation coefficient, [1/km].

Engineers usually think of attenuation in terms of decibels (treated more extensively in Chapter 10); therefore, the equation may be rewritten using $\alpha = 0.4343 \, \alpha'$, from conversion of base e to base 10.

$$P(L) = P_0 \cdot 10^{-\alpha L}$$
Eq. 3.11
$$\log P(L) = -\alpha L / \log 10 + \log P_0$$
$$\alpha = \alpha_{scattering} + \alpha_{absorption} + \alpha_{bending}$$

Where:

α = Fiber loss, [dB/km].

Attenuation in optical fiber is caused by several intrinsic and extrinsic factors. Two intrinsic factors are scattering and absorption. The most common form of scattering, Rayleigh Scattering (see Figure 3.13), is caused by microscopic non-uniformities in the optical fiber. These non-uniformities cause rays of light to partially scatter as they travel along the fiber; thus, some light energy is lost. Rayleigh scattering represents the strongest attenuation mechanism in most modern optical fibers; nearly 90% of the total attenuation can be attributed to it. It becomes important when the size of the structures in the glass itself are comparable in size to the wavelengths of light traveling through the glass. Thus, long wavelengths are less affected than short wavelengths. The attenuation coefficient (α)

decreases as the wavelength (λ) increases and is proportional to λ^{-4}. Rayleigh scattering increases sharply at short wavelengths. (Rayleigh scattering causes the sky to be blue. Only the shortest visible wavelengths, blue, are significantly scattered by air molecules.)

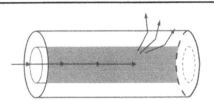

Figure 3.13:
Scattering

Absorption, illustrated in Figure 3.14, can be caused by the molecular structure of the material, impurities in the fiber such as metal ions, OH⁻ ions (water), and atomic defects such as unwanted oxidized elements in the glass composition. These impurities absorb the optical energy and dissipate it as a small amount of heat. As this energy dissipates, the light becomes dimmer. At 1.25 and 1.39 μm wavelengths, optical loss occurs because of the presence of OH⁻ ions in the fiber. Above a wavelength of 1.7 μm, glass starts absorbing light energy due to the molecular resonance of the SiO_2 molecule.

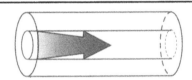

Figure 3.14:
Absorption

Extrinsic causes of attenuation include cable manufacturing stresses, environmental effects, and physical bends in the fiber. Physical bends break down into two categories: microbending and macrobending (see Figure 3.15). Microbending is the result of microscopic imperfections in the geometry of the fiber. These imperfections could be rotational asymmetry, changes of the core diameter, rough boundaries between the core and cladding, a result of the manufacturing process itself, or mechanical stress, pressure, tension, or twisting. Macrobending describes fiber curvatures with diameters on the order of centimeters. The loss of optical power is the result of less-than-total reflection at the core-to-cladding boundary. In single-mode fiber, the fundamental mode is partially converted to a radiating mode due to the bends in the fiber. Because of the increase in mode field diameter with wavelength, in a given single-mode fiber, macrobending loss will be higher at longer wavelengths. Bending loss is usually unnoticeable if the radius of the bend is larger than 10 cm.

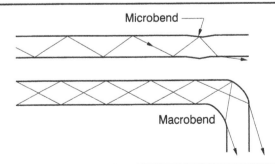

Microbend

Macrobend

Figure 3.15: Bending

The amount of attenuation caused by an optical fiber is primarily determined by its length and the wavelength of the light traveling through the fiber. There are also many secondary and tertiary factors that contribute.

Figure 3.16: Optical
Loss versus
Wavelength

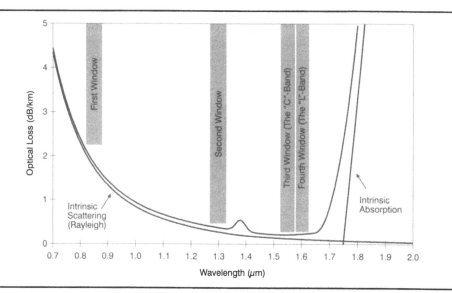

Figure 3.16 shows the loss per unit length of a typical modern optical fiber. The plot covers wavelengths from 0.7 μm to 2.0 μm. As a point of reference, the human eye sees light in the range from 0.4 μm (blue) to 0.7 μm (red). Most modern fiber optic transmission takes place at wavelengths longer than red, in the infrared region. There are five important fiber optic wavelength regions: 780 nm, 850 nm, 1310 nm, 1550 nm, and 1625 nm. These particular wavelengths were chosen because the loss of the fiber is lowest at these wavelengths. There are three primary mechanisms that influence the fiber's loss at a given wavelength. At shorter wavelengths, Rayleigh scattering is important, varying proportionally to λ^{-4}. At longer wavelengths, absorption becomes dominant as the molecules in the glass start to resonate. In between, absorption by impurities is important. Refer back to Figure 1.3 to see the absorption peaks common in older fiber types.

FIBER RIN

Optical fiber itself can generate noise in a transmission system. This noise, usually only a concern for analog systems, is called fiber-induced RIN (relative intensity noise) or double Rayleigh scattering.

Figure 3.17: Fiber-
induced RIN

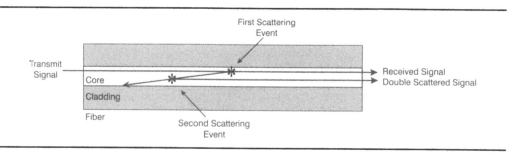

Figure 3.17 illustrates fiber RIN. Some light always scatters (e.g. Rayleigh) in fiber, contributing to attenuation. Some of this scattered light travels backwards through the fiber where it again scatters and changes direction, emerging from the fiber together with the desired signal. A large separation between the scattering events can cause a long time delay between the signal and the double-scattered light (much longer than data pulses). The beating (constructive and destructive interference) between the signal and the multiple scattered light causes noise and distortion at the receive end of the fiber.

MULTIMODE DISPERSION

Multimode dispersion describes the pulse broadening in multimode fibers caused by different modes traveling at different speeds through the fiber. This form of dispersion is sometimes called modal dispersion because it is characteristic of multimode fiber only. Multimode dispersion can be reduced in three ways:

1. Use a smaller core diameter fiber to reduce the number of modes traveling through the fiber.

2. Use graded-index fiber. As described earlier, graded-index fiber uses different refractive indices within the fiber to move the modes along so that they arrive at the end of the fiber together.

3. Use single-mode fiber. Under most circumstances, this option eliminates multimode dispersion altogether.

To qualify number three, single-mode fiber is only single-mode at wavelengths longer than the cutoff wavelength. For a typical 1310 nm single-mode fiber (9 μm core diameter), this cutoff wavelength is between 1150 nm and 1200 nm. At wavelengths below this threshold, the single-mode fiber will become dual mode, then tri-mode and so on. Increasingly, 9 μm single-mode fiber is being used with short-wavelength lasers at 780 nm to 850 nm. At these wavelengths, 9 μm core fiber is actually dual-mode. There is a specialty size single-mode fiber with a 5 μm core diameter, but it is rarely used because of its cost and the difficulty coupling a light source to such a small core size. The point is, multimode dispersion is also a factor in short wavelength systems that use 9 μm core diameter single-mode fiber. When multimode dispersion is present, it usually dominates to the point that other types of dispersion can typically be ignored.

CHROMATIC DISPERSION

Chromatic dispersion represents the fact that different colors or wavelengths travel at different speeds, even within the same mode. Chromatic dispersion is the result of material dispersion, waveguide dispersion, or profile dispersion. Figure 3.18 below shows chromatic dispersion along with its key components, waveguide dispersion and material dispersion. In this example, chromatic dispersion goes to zero at a wavelength near 1550 nm. This is characteristic of dispersion-shifted fiber. Standard fiber, single-mode or multimode, has zero dispersion at a wavelength of 1310 nm.

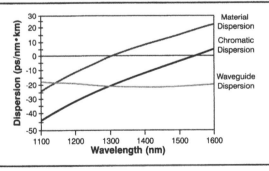

Figure 3.18:
Chromatic Dispersion

If you can operate a fiber at the zero-dispersion wavelength with a monochromatic light source, the bandwidth of the fiber will be very large. Figure 3.19 shows the bandwidth-distance product for a hypothetical single-mode fiber. The x-axis is the center wavelength for the light source. Three curves are shown for light sources with FWHM (Full Width Half Maximum) values of 2 nm, 5 nm and 10 nm. FWHM is the width of the spectral emission at the 50% amplitude points.

Figure 3.19: Single-mode Fiber Bandwidth

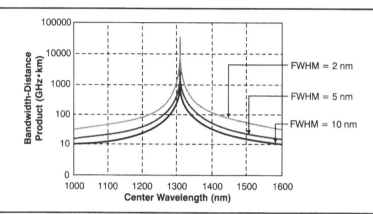

For the most narrow source shown, FWHM = 2 nm, the bandwidth-distance product for the fiber is over 30,000 GHz•km at a center wavelength of 1310 nm, the fiber's zero-dispersion wavelength. As the center wavelength moves even a few nanometers from 1310 nm, the fiber's bandwidth-distance product drops dramatically. At a center wavelength of 1310 nm, the fiber's bandwidth-distance product has dropped by a factor of 30, to 1,000 GHz•km. For wider optical sources, the bandwidth is even lower, as can be seen in the plot. If one used a very narrow light source tuned exactly for the fiber's zero-dispersion wavelength, the peak would be much higher than those shown in the plot.

Material Dispersion

Different wavelengths travel at different velocities through a fiber, even in the same mode. We know that the index of refraction (n) is given as:

Eq. 3.12 $$n = \frac{c}{v}$$

Where:

c = The speed of light in a vacuum.
v = The speed of the same wavelength in the material.

Each wavelength travels at a different speed through the material. This changes the value of v in the equation at each wavelength. Dispersion from this phenomenon is called material dispersion. Two factors determine the amount of material dispersion that will occur.

1. The range of light wavelengths injected into the fiber: a source emits several wavelengths rather than a single wavelength. This is the spectral width of a source. LED's have a much wider spectral width (about 35-170 nm) than lasers (< 5 nm).

2. The center operating wavelength of the source: around 850 nm, longer wavelengths (red) travel faster than shorter wavelengths (blue). However, at 1550 nm, the situation is reversed, and the shorter blue wavelengths travel faster than the longer red ones. The crossover where the wavelengths travel at the same speed occurs around 1310 nm, the zero-dispersion wavelength.

Material dispersion greatly affects single-mode fibers. In multimode fibers, multimode dispersion usually supersedes material dispersion in terms of its impact on the system. Figure 3.20 shows the refractive index versus wavelength for pure fused silica core fiber. The group refractive index, also called group index, is the speed of light in a vacuum (c) divided by the group velocity of the mode. This group velocity is the reciprocal of the rate change of the phase constant with respect to the angular frequency. Keep in mind that this relationship varies according to the composition of the fiber core.

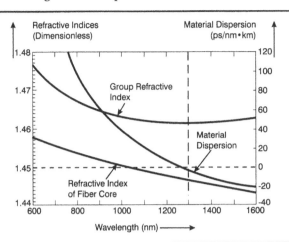

Figure 3.20: Material Dispersion

Waveguide Dispersion

Like material dispersion, waveguide dispersion is a greater concern for single-mode fibers than for multimode fibers. Waveguide dispersion occurs because optical energy travels in both the core and the cladding at slightly different speeds. This is a result of the difference in the indices of refraction between the core and the cladding. Practical single-mode fiber is designed so that material dispersion and waveguide dispersion cancel one another at the wavelength of interest.

Profile Dispersion

The refractive indices of both the core and the cladding affect the group velocity in a fiber. These differences are usually stated in terms of the refractive index profile. This describes the value of the refractive index as a function of distance from the optical axis along an optical fiber diameter. Profile dispersion is caused by the different wavelength dependencies of the refractive indices of the core and cladding. These differences are caused by the different materials involved. This parameter is more important in multimode fibers than in single-mode fibers because the profile can be optimized for only one wavelength.

POLARIZATION MODE DISPERSION

Polarization mode dispersion (PMD) is another complex optical effect that occurs in single-mode optical fibers. As discussed in Chapter 2, most single-mode fibers support two perpendicular polarizations of the original transmitted signal. If a fiber was perfect, i.e. perfectly round and free from all stresses, both polarization modes would propagate at exactly the same speed, resulting in zero PMD. In actual fibers, the two perpendicular polarizations may travel at different speeds and consequently arrive at the end of the fiber at different times. The fiber is said to have a fast axis, and a slow axis. The difference in arrival times, normalized with length, is known as PMD $(ps/\sqrt{km}\;)$.

When light is injected into an optical fiber, the light usually splits into the two different polarization planes and each polarization travels down the fiber. PMD occurs in optical fiber because light in each of the polarization planes within an optical fiber can travel at slightly different speeds, leading to a distortion of the pulse shape. Figure 3.21 shows how differences in velocity between two polarizations of light traveling down a fiber can lead to pulse distortion or broadening as the light on the fast axis races ahead of light on the slow axis. PMD is an important factor in high bit rate transmissions — typically transmission rates greater than 10 Gbit/s.

Figure 3.21: Pulse Broadening Due to PMD

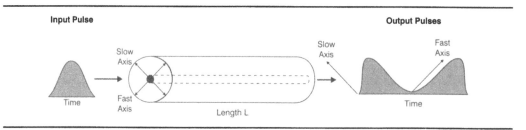

The newest and least studied aspect of optical fiber, PMD, became a serious issue when data rates reached OC-192 levels. PMD varies with time while most other fiber distortions and nonlinearities are relatively stable over time. Various techniques exist to correct for PMD. Optical techniques usually involve applying stress to a fiber at multiple points. Electrical techniques also exist to correct PMD distortion after the photodetector and amplifier.

DISPERSION MANAGEMENT

Dispersion and its adverse affects on fiber optic signal transmission have been covered in detail in this chapter. Dispersion management refers to the process designing the fiber and compensating elements in the transmission path to keep the total accumulated dispersion to a small number. Typically in a modern fiber optics system, dispersion compensating elements are placed approximately every 100 km.

An optical pulse subjected to positive dispersion will see its longer wavelength components travel slower than its shorter counterparts, whereas an optical pulse subjected to negative dispersion will see its longer wavelength components travel faster than its shorter counterparts. This last property is important, as a light pulse that has undergone a certain amount of negative dispersion can essentially correct itself by traveling through a medium that has an equivalent amount of positive dispersion. Dispersion-compensating modules perform this function in fiber optic links by introducing, in-line, an opposite dispersion to that of the transmission fiber.

Dispersion management is especially critical in WDM and DWDM long-haul applications. Wavelength-division multiplexing technology easily expands the transmission capacity of optical fiber. One recent technique, using a dispersion slope compensator (DSC), shows great promise for long distance, high bit rate optical communications systems.

Figure 3.22 illustrates the DSC configuration in this example. In this case, an arrayed waveguide grating (AWG)-based DSC was used at the send and receive ends. An AWG is a type of planar lightwave circuit (PLC), an optical circuit laid out on a silicon wafer that integrates multiple optical functions. In the AWG these functions include dividing the light into more than 100 separate fiber gratings used to filter specific wavelengths of light. At the receiving end, a double-pass style ultra-narrow bandwidth optical filter was used to select the desired WDM channel or wavelength. As the chromatic dispersion of each signal was equalized by the DSC, no equalization fiber was needed. The optical attenuators were used to adjust the loss difference between the channels.

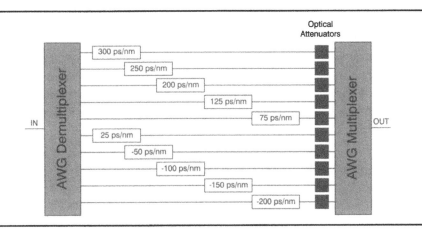

Figure 3.22:
Dispersion Slope
Compensator Using
AWG's

Figures 3.23 and 3.24 show the accumulated chromatic dispersion along the transmission distance with and without the DSC illustrated in Figure 3.22. The DSC greatly reduced the difference of the chromatic dispersion accumulation between the channels.

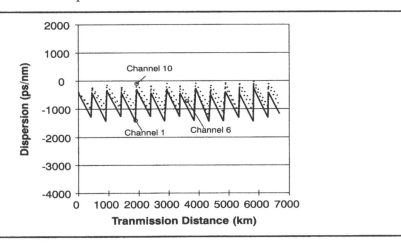

Figure 3.23:
Accumulated
Chromatic Dispersion
Using DSC

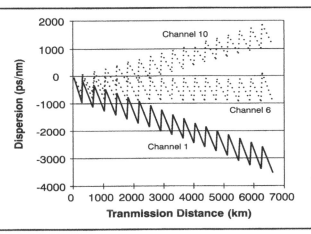

Figure 3.24:
Accumulated
Chromatic Dispersion
Without DSC

FIBER NONLINEARITIES

As optical fiber data rates, transmission lengths, number of wavelengths, and optical power levels increased, a host of nonlinear fiber effects, just laboratory curiosities a few years earlier, suddenly became very important. In the early days of fiber, one had to worry most about fiber attenuation and, to some extent, fiber dispersion. As the fiber performance envelope stretched, dispersion became more important, but this effect is generally well understood and can be dealt with using a variety of dispersion avoidance and cancellation techniques. What is less well known are the various optical fiber nonlinearities that have previously not been seen in field deployments other than specialized applications such as undersea installations. Fiber nonlinearities that now must be considered in designing state-of-the-art fiber optic systems include stimulated Brillouin scattering (SBS), stimulated Raman scattering (SRS), four wave mixing (FWM), self-phase modulation (SPM), cross-phase modulation (XPM), and intermodulation (mixing).

Why are all of these nonlinearities now so important? Because they now represent the fundamental limiting mechanisms to the amount of data that can be transmitted on a single optical fiber. System designers must be aware of these limitations and the steps that can be taken to minimize the detrimental effects of fiber nonlinearities.

Fiber nonlinearities arise from two basic mechanisms. The first, and most serious, is that the refractive index of glass is dependent on the optical power going through the material. The general equation for the refractive index of the core in an optical fiber is:

$$\text{Eq. 3.13} \qquad n = n_0 + n_2 \cdot P/A_{eff}$$

Where:

n_0 = The refractive index of the fiber core at low optical power levels.
n_2 = The nonlinear refractive index coefficient (2.35×10^{-20} m^2/W for silica).
P = The optical power in Watts.
A_{eff} = The effective area of the fiber core in square meters.

The equation shows that nonlinearities produced by refractive index power dependence are eliminated by minimizing the amount of power, P, launched and maximizing the effective area of the fiber, A_{eff}. Obviously, minimizing P is counter to the current trend. We can, however, maximize A_{eff} without detrimental effects. Maximizing this effective area remains the focus of the latest fiber designs.

Figure 3.25: Silica Refractive Index versus Optical Power

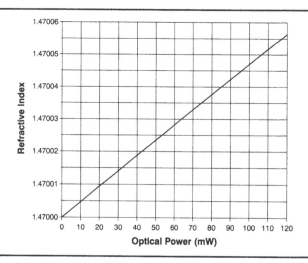

Figure 3.25 shows the relationship of the refractive index of silica versus optical power. It can be seen that the magnitude of the change in refractive index is relatively small. This only becomes important because the interaction length in a real fiber optic system can be hundreds of kilometers. The power dependent refractive index of silica gives rise to the SPM, XPM, FWM, and intermodulation (mixing) nonlinearities. Scattering phenomena, the second set of mechanisms for generating nonlinearities, give rise to SBS and SRS.

Stimulated Brillouin Scattering

Stimulated Brillouin Scattering (SBS) imposes an upper limit on the amount of optical power that can be usefully launched into an optical fiber. The SBS effect has a threshold optical power. When the SBS threshold is exceeded, a significant fraction of the transmitted light is redirected back toward the transmitter. This results in a saturation of optical power that reaches the receiver, as well as problems associated with backreflection in the optical signals. The SBS process also introduces significant noise into the system, resulting in degraded bit error rate (BER) performance. Controlling SBS is particularly important in high-speed transmission systems that employ external modulators and continuous wave (CW) laser sources. It is also of vital importance to the transmission of 1550 nm CATV signals since these transmitters often have the very characteristics that trigger the SBS effect.

Time-varying electric fields within a fiber interact with the acoustic vibrational modes of the fiber material which in turn scatter the incident light. This is known as Brillouin scattering. When the source of the high intensity electric fields is the incident lightwave, the effect is known as SBS. The high power incident lightwave actually causes the refractive index of the fiber to vary periodically causing backreflection similar to the effect of Bragg gratings, a type of optical filter incorporated into optical fiber. As the input optical level increases beyond the SBS threshold, an increasingly larger portion of the light is backscattered, creating an upper limit to the power levels that can be carried over the fiber. Figure 3.26 illustrates this phenomenon. As the launch power is increased above the threshold, there is a dramatic increase in the amount of backscattered light. The precise threshold for the onset of the SBS effect depends on a number of system parameters including wavelength (the threshold is lower at 1550 nm than 1310 nm) and linewidth of the transmitter. Values of +8 to +10 dBm are typical for direct modulated optical sources operating at 1550 nm over standard single-mode fiber.

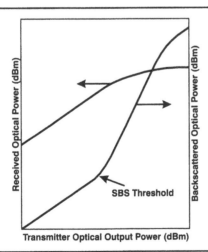

Figure 3.26: SBS Threshold Effects

The SBS threshold is strongly dependent on the linewidth of the optical source with narrow linewidth sources having considerably lower SBS thresholds. Extremely narrow linewidth lasers (e.g. less than 10 MHz wide), often used in conjunction with external modulators, can have SBS thresholds of +4 to +6 dBm at 1550 nm. The SBS threshold increases proportionally as the optical source linewidth increases as shown in Figure 3.27. Broadening the effective spectral width of the optical source minimizes SBS.

One approach for broadening linewidth involves externally modulating the transmitter, while spreading out the linewidth by adding a very small AC modulation signal to the DC current source used to drive the laser itself. This broadens the spectral linewidth of the transmitter and increases the threshold for the onset of SBS. This option also increases the dispersion susceptibility of the transmitter, primarily a concern when operating at 1550 nm over non dispersion-shifted single-mode fiber. Practical implementations of SBS suppression circuitry based on laser drive dithering can increase the SBS threshold by 5 dB.

Figure 3.27: SBS Threshold versus Source Linewidth

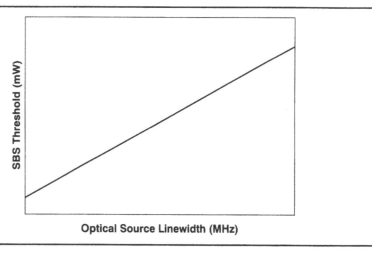

Another common means for increasing the SBS threshold is to phase dither the output of the external modulator. In this case, a high frequency signal, usually twice that of the highest frequency being transmitted, is imposed onto both output legs of the external modulator. This modulates the phase of the light, effectively spreading out the spectral width. Figure 3.28 shows the optical spectra of an VSB/AM transmitter without phase dithering. The central carrier exceeds the SBS threshold, causing serious system degradation. (See Chapter 7 for details of external modulator operation.)

Figure 3.28: Optical Spectrum without Phase Modulation

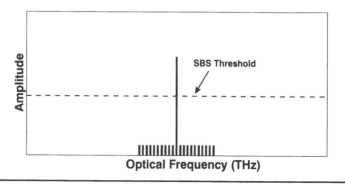

In Figure 3.29, a high frequency dither signal has been applied to the phase modulation input of the external modulator. It can be seen that all of the lines are now comfortably below the SBS threshold. This technique can raise the SBS threshold optical power by about 10 dB.

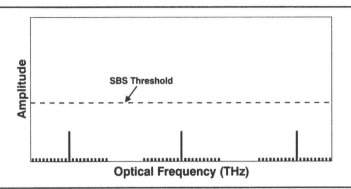

Figure 3.29: Optical Spectrum with Phase Modulation

The other factor to consider is that the SBS threshold decreases when a number of EDFA's are in the signal path. The SBS threshold for a system containing N amplifiers is the threshold without amplifiers in mW divided by N. This can result in very low SBS thresholds that can seriously impair system performance.

Stimulated Raman Scattering

Stimulated Raman scattering (SRS) is much less of a problem than SBS. Its threshold is close to 1 Watt, nearly a thousand times higher than SBS. But real systems are being deployed with EDFA's having optical output powers of 500 mW (+27 dBm), and this will only go higher. A fiber optic link that includes three such optical amplifiers will reach this limit since the limit drops proportionally by the number of optical amplifiers in series. SRS can cause scattering like SBS, but usually the effect first seen is that the shorter wavelength channels are robbed of power, and that power feeds the longer wavelength channels. This is similar to the operation of EDFA's where a 980 nm pump wavelength provides the energy that amplifies the signals in the longer wavelength, 1550 nm, region. Currently, EDFA's use a special erbium-doped fiber to provide this gain mechanism.

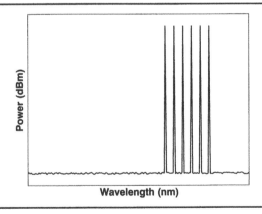

Figure 3.30: Six Channel DWDM Transmitted Optical Spectrum

Figure 3.30 shows the typical transmit spectrum or a six-channel DWDM system. Note that all of the six wavelengths have identical amplitudes. These signals are all in the 1550 nm window. In Figure 3.31, it can be seen that the short wavelength channels have a much smaller amplitude compared to the longer wavelength channels. This is the SRS effect.

Plain silica fiber can provide similar gain using the Raman gain mechanism. Raman amplifiers are only now becoming mainstream additions to long-haul telecommunications system. They will be covered more extensively in Chapter 7.

Figure 3.31: SRS Effect on Six Channel DWDM Transmitted Optical Spectrum

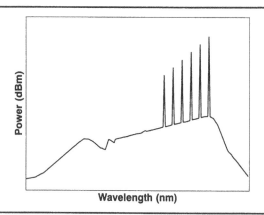

Four Wave Mixing

Four wave mixing (FWM) usually only shows up in fiber optic transmission systems that carry lots of simultaneous wavelengths, such as DWDM systems. Caused by the nonlinear nature of the refractive index of the optical fiber itself, the FWM effect is similar to composite triple beat (CTB) distortion observed in CATV systems. CTB is also caused by nonlinearity, this time in the electrical amplifier chain or one of the optical components, usually the laser. CTB, like FWM, is classified as a third-order distortion phenomenon. Third-order distortion mechanisms generate third-order harmonics in systems with one channel. In multichannel systems, third-order mechanisms generate third-order harmonics and a gamut of cross products. These cross products cause the most problems since they often fall near or on top of the desired signals.

Consider a simple three-wavelength (λ_1, λ_2, and λ_3) system that is experiencing FWM distortion. In this simple system, nine cross products are generated near λ_1, λ_2, and λ_3 that involve two or more of the original wavelengths. Note that there are additional products generated, but they fall well away from the original input wavelengths. Let us assume that the input wavelengths are λ_1 = 1551.72 nm, λ_2 = 1552.52 nm, and λ_3 = 1553.32 nm. The interfering wavelengths that are of most concern in our hypothetical three wavelength system are:

$\lambda_1 + \lambda_2 - \lambda_3$ = 1550.92 nm	$\lambda_1 - \lambda_2 + \lambda_3$ = 1552.52 nm	$\lambda_2 + \lambda_3 - \lambda_1$ = 1554.12 nm
$2\lambda_1 - \lambda_2$ = 1550.92 nm	$2\lambda_1 - \lambda_3$ = 1550.12 nm	$2\lambda_2 - \lambda_1$ = 1553.32 nm
$2\lambda_2 - \lambda_3$ = 1551.72 nm	$2\lambda_3 - \lambda_1$ = 1554.92 nm	$2\lambda_3 - \lambda_2$ = 1554.12 nm

It can be seen that three of the interfering products fall right on top of the original three signals. The remaining six products fall outside of the original three signals. These six can be optically filtered out. Because the three interfering products that fall on top of the original signals are mixed together, they cannot be removed by any means. Figure 3.32 shows the results graphically. The three tall solid bars are the three original signals. The shorter cross-hatched bars represent the nine interfering products. The number of interfering products increases as $\frac{1}{2} \cdot (N^3 - N^2)$ where N is the number of signals.

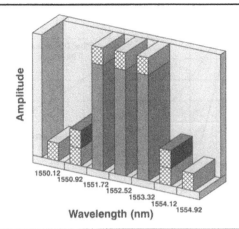

Figure 3.32: FWM Products for a Three-wavelength System

Figure 3.33 shows that the number of interfering products rapidly becomes a very large number. Since there is no way to eliminate products that fall on top of the original signals, the only hope is to prevent them from forming in the first place.

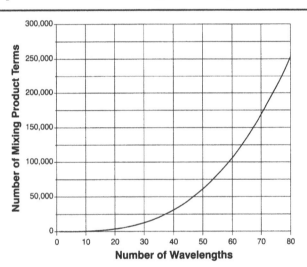

Figure 3.33: FWM Products versus Channel Count

Two factors strongly influence the magnitude of the FWM products, referred to as the FWM mixing efficiency. The first factor is the channel spacing; mixing efficiency increases dramatically as the channel spacing becomes closer. Fiber dispersion is the second factor, and mixing efficiency is inversely proportional to the fiber dispersion, being strongest at the zero-dispersion point. In all cases, the FWM mixing efficiency is expressed in dB, and more negative values are better since they indicate a lower mixing efficiency.

Figure 3.34 shows the magnitude of FWM mixing efficiency versus fiber dispersion and channel spacing. If a system design uses NDSF with dispersion of 17 ps/nm•km and the minimum recommended International Telecommunication Union (ITU) DWDM spacing of 0.8 nm, then the mixing efficiency is about -48 dB and will have little impact. On the other hand, if a system design uses DSF with a dispersion of 1 ps/nm•km and a non-standard spacing of 0.4 nm, then the mixing efficiency becomes -12 dB and will have a severe impact on system performance, perhaps making recovery of the transmitted signal impossible. The data presented in Figure 3.34 is for a given optical power level, fiber length, wavelength and so on. The magnitude of the mixing efficiency will vary widely as these parameters vary. The data presented is intended to illustrate the principles only.

Figure 3.34: FWM
Mixing Efficiency in
Single-mode Fibers

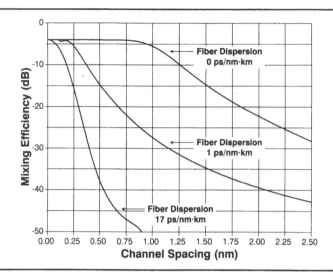

Self-phase Modulation

Figure 3.35 illustrates self-phase modulation. Like FWM, self-phase modulation (SPM) is due to the power dependency of the refractive index of the fiber core. It interacts with the chromatic dispersion in the fiber to change the rate at which the pulse broadens as it travels down the fiber. Whereas increasing the fiber dispersion will reduce the impact of FWM, it will increase the impact of SPM. As an optical pulse travels down the fiber, the leading edge of the pulse causes the refractive index of the fiber to rise, resulting in a blue shift. The falling edge of the pulse decreases the refractive index of the fiber causing a red shift. These red and blue shifts introduce a frequency chirp on each edge which interacts with the fiber's dispersion to broaden the pulse.

Figure 3.35: Effects of
SPM on a Pulse

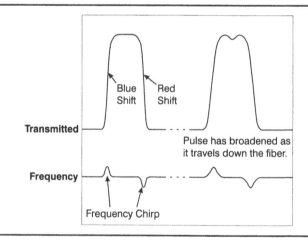

Cross-phase Modulation

Cross-phase modulation is very similar to SPM except that it involves two pulses of light, whereas SPM needs only one pulse. In XPM, two pulses travel down the fiber, each changing the refractive index as the optical power varies. If these two pulses happen to overlap, they will introduce distortion into the other pulses through XPM. Unlike, SPM, fiber dispersion has little impact on XPM. Increasing the fiber effective area will reduce XPM and all other fiber nonlinearities.

Intermodulation (Mixing)

Intermodulation is fairly similar to SPM and XPM. Consider the case where two laser light sources are transmitting light through the fiber. Again as the optical power in each light wave peaks and drops, the refractive index of the fiber changes accordingly. Now the two different light sources have different frequencies f_1 and f_2. As the refractive index changes in concert with frequencies f_1 and f_2, new frequencies, $2 \cdot f_1 - f_2$ and $2 \cdot f_2 - f_1$, appear. This is similar in many ways to the FWM nonlinearity.

Summary of Fiber Nonlinearities

Below is a quick summary of the important characteristics of all these fiber nonlinearities.

SBS:

- SBS is caused by an interaction between the incident lightwave and the acoustical vibration modes in the fiber.
- The SBS threshold is directly proportional to the fiber area. Dispersion-shifted fibers have smaller areas, thus a lower threshold.
- Fiber designs with larger effective areas have a higher SBS threshold.
- Threshold is directly proportional to the laser linewidth. Direct modulation, dithering of a CW laser or phase modulation of an external modulator all increase the effective linewidth, raising the threshold.
- Narrow linewidth laser sources in the 1550 nm region without countermeasures can encounter SBS at optical powers of only +5 dBm (3 mW).
- Countermeasures (dithering of the CW laser drive and phase modulating the output) can increase the SBS threshold in the 1550 nm region to about +16 dBm (40 mW).
- SBS limits the amount of light that reaches the receiver.
- Above the SBS threshold, backscattered light increases dramatically as does noise reaching the receiver.
- The SBS threshold is lower at longer wavelengths.
- For a system consisting of a chain of N optical amplifiers, the SBS threshold, in milliwatts, will drop by a factor of N.

SRS:

- The SRS threshold power is about +30 dBm (1 Watt).
- SRS limits the amount of light that reaches the receiver above the threshold.
- Below the SRS threshold, the SRS effect can rob power from shorter wavelength channels and feed that power to longer wavelength channels.
- For a system consisting of a chain of N optical amplifiers, the SRS threshold, in milliwatts, will drop by a factor of N.
- Fiber designs with larger effective areas have a higher SRS threshold.

FWM:

- FWM is a phenomenon that arises from the nonlinearity of the refractive index of the optical fiber.
- FWM is a third-order distortion mechanism. It is very similar to CTB (composite triple beat) distortion in the CATV realm.
- FWM becomes worse as the fiber dispersion drops. It is worst at the zero-dispersion point. Higher chromatic dispersion results in less FWM.
- FWM is worst in WDM channel designs where the spacing is equal. (Equal channel spacing is, unfortunately, the case in standardized DWDM designs.)
- FWM becomes worse as wavelengths are spaced closer together.
- Fiber designs with larger effective areas exhibit less FWM nonlinearity.

SPM:

- SPM causes a frequency chirp on the rising and falling edges of an optical pulse, broadening the pulse.
- SPM effects a single light pulse traveling down the fiber.
- SPM acts along with chromatic dispersion to broaden pulses.
- Higher chromatic dispersion results in more SPM.
- Fiber designs with larger effective areas have a higher SPM threshold.

XPM:

- XPM causes multiple pulses traveling down the fiber to interact through their mutual effect on the refractive index of the fiber.
- XPM causes pulses to become distorted as they interact.
- Fiber dispersion has little effect on XPM.
- Fiber designs with larger effective areas have a higher XPM threshold.

Intermodulation (Mixing):

- Intermodulation is similar to XPM except that it causes new frequency components to appear that are cross products of the original frequencies.

SELECTING FIBER TYPES

This section offers a guide to selecting the correct type of fiber for a given application. Several factors must be considered:

- Fiber core/cladding size (e.g., 50/125 μm):
- Fiber material and construction (e.g., glass core/glass cladding)
- Fiber attenuation measured in dB/km
- Fiber bandwidth-distance product measured in MHz•km
- Environmental considerations (e.g., temperature)

The first number listed (e.g., 50 or 62.5) represents the fiber core diameter in microns. The number after the slash is the cladding diameter. This is the key dimension when picking a fiber optic connector. For instance, 50/125 μm and 62.5/125 μm size fibers can use the same size connector. A connector with a larger hole would be required for 100/140 μm fiber. Figure 3.36 illustrates the most popular fiber sizes. By studying the figure, one can get a visual feel for the relative core and cladding size that is useful for identifying an unknown fiber's core and cladding using a microscope.

Figure 3.36: Popular Fiber Sizes — Magnified Drawings

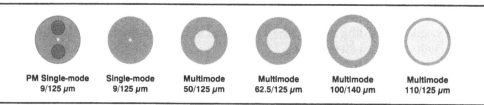

Most fiber optic applications utilize fibers made from glass cores and glass cladding. This combination yields excellent performance at a reasonable cost. Other combinations include plastic clad silica (plastic cladding/silica core, called PCS) and all-plastic fiber. PCS fiber is used mainly in applications where high nuclear radiation levels may be present. It is also widely used for imaging fibers because of its superior transmission at visible wavelengths. All-plastic fibers can cost less, but they are typically limited to very short distances (a few meters) because attenuation is very high and bandwidth-distance product is low. Figure 3.37 has three sections labeled Attenuation, Bandwidth-Distance Products, and Power, that gives a quick visual comparison between three important fiber parameters.

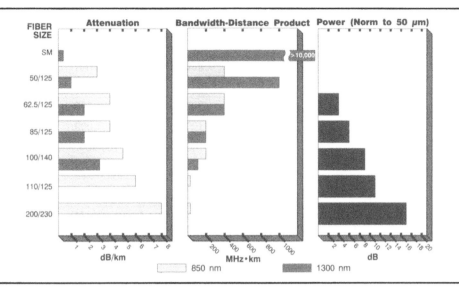

Figure 3.37: Comparison of Optical Fiber Types

The attenuation section lists performance by different size fibers. Smaller numbers indicate the better fiber. Bandwidth-distance products tell how much information can be carried per second on a single fiber. Higher numbers indicate the greater information carrying capacity. Power refers to the relative optical power that can be coupled into an optical fiber using an LED. In this graph, 50/125 μm fiber is the reference. With 62.5/125 μm fiber, one can typically couple 4.5 dB more power into the fiber. That amounts to 181% more power compared to 50/125 μm fiber. Single-mode fiber, if shown, would be about -13 dB relative to 50/125 μm fiber. As is the case with most engineering trade-offs, you do not get something for nothing. Fibers that are easy to couple into a lot of power tend to have poor bandwidth-distance products and attenuation rates.

Fiber loss, an important parameter for selecting optical fiber, depends heavily on the operating wavelength. Practical fibers have the lowest loss at 1550 nm and the highest loss at 780 nm. Fiber bandwidth is another critical parameter, determining the maximum rate at which data can be transmitted. This commonly used parameter only gives the bandwidth limitation due to multimode dispersion. Material dispersion may be more limiting, especially at shorter wavelengths. The discussion earlier in this chapter gives a better description of the dispersion mechanisms that must be considered in a full systems analysis. Table 3.2 shows the typical fiber optical loss at four key fiber optic wavelengths. The data is taken from a number of fiber manufacturers. Many manufacturers offer multiple levels of performance for each fiber size. The table lists the range of values seen with mainstream fibers.

Fiber		Optical Loss (dB/km)			
Size	Type	780 nm	850 nm	1310 nm	1550 nm
9/125 μm	SM	~~~	~~~	0.5-0.8	0.2-0.3
50/125 μm		4.0-8.0	3.0-7.0	1.0-3.0	1.0-3.0
62.5/125 μm		4.0-8.0	3.0-7.0	1.0-4.0	1.0-4.0
100/140 μm	MM	4.5-8.0	3.5-7.0	1.5-5.0	1.5-5.0
110/125 μm		~~~	15.0	~~~	~~~
200/230 μm		~~~	12.0	~~~	~~~

Table 3.2: Typical Fiber Loss

Table 3.3 presents the bandwidth-distance product of the common fiber sizes at four fiber optic wavelengths. Table 3.4 shows other miscellaneous fiber parameters.

Table 3.3: Typical
Fiber Bandwidth

Fiber		Bandwidth-Distance Product (MHz•km)			
Size	**Type**	**780 nm**	**850 nm**	**1310 nm**	**1550 nm**
9/125 μm	SM	<800	2,000	20,000+	4,000-20,000+
50/125 μm		150-700	200-800	400-1,500	300-1,500
62.5/125 μm		100-400	100-400	200-1,000	150-500
100/140 μm	MM	100-400	100-400	100-400	10-300
110/125 μm		---	17	---	---
200/230 μm		---	17	---	---

Fiber Size	Numerical Aperture	Temperature Range	Min. Bend Radius
9/125 μm	---	-60 to +85°C	12 mm
50/125 μm	0.20	-60 to +85°C	12 mm
62.5/125 μm	0.275	-60 to +85°C	12 mm
100/140 μm	0.29	-60 to +85°C	12 mm
110/125 μm	0.37	-65 to +125°C	15 mm
200/230 μm	0.37	-65 to +125°C	16 mm

There are also environmental factors to consider in selecting a fiber. Temperature is often the most demanding parameter. Surprisingly, most optical fibers perform worse at low temperatures because the cable material becomes stiff and puts the fiber under stress.

Some generalizations that can be made about fiber types are as follows:

- Larger core size fibers are generally more expensive.
- Larger core size fibers have higher loss (attenuation) per unit distance.
- Larger core size fibers have lower bandwidth.
- Larger core size fibers allow lower cost light sources to be used.
- Larger core size fibers typically use lower cost connectors.
- Larger core size fibers typically yield the lowest system cost (for short distance systems).

The first three items tend to drive the selection to smaller fibers while the last three items drive the selection to larger fibers. Many factors will dictate the choice of fiber type. Transmission bandwidth, maximizing distance between repeaters/amplifiers, cost of splicing or connectorizing, tolerance of temperature fluctuations, strength and flexibility are just some of these factors. In most applications, cabling the fiber is required to meet the application requirements. (Fiber optic cables are discussed in detail in the next chapter.)

It is impossible to say which fiber type is best without examining the specific problem to be solved. There are applications where 200/230 μm fiber is the best choice and others where single-mode fiber is the optimum choice. Which fiber is best is akin to asking which car is best. If economy is the only consideration, one might choose a Hyundai. If speed is the goal, one might pick a Porsche. If luxury is the goal, then a Cadillac might be the best choice. Despite the array of fiber types, only a few are important. These are:

9/125 μm (SM): Widely used for high data rate and long distance applications.

62.5/125 μm (MM): Very popular in most commercial applications; it has wide uses with low- to-moderate-speed data links and video links.

50/125 μm (MM): This fiber type is mainly used by military customers, but it is experiencing a small comeback in commercial applications.

100/140 μm (MM): Once a very popular size, there are only a few remaining applications. The only major application is use in aircraft.

Chapter Summary

- Optical fiber evolved from early, high-loss designs, to low-loss 1550 nm fiber types that can transmit DWDM signals.
- Inside vapor deposition and outside vapor deposition are the two predominant methods for manufacturing optical fiber.
- The operation of optical fiber is based on the principle of total internal reflection which occurs because light travels at different speeds in different materials.
- As light passes from one medium to another medium with a different index of refraction, the light is bent, or refracted.
- The core of an optical fiber has a higher index of refraction than the cladding, allowing for total internal reflection.
- Multimode fiber and single-mode fiber are the two basic types of optical fiber.
- A logarithmic relationship between the optical output power of the transmitter and the optical input power of the receiver in a fiber optic system expresses attenuation.
- Scattering, absorption, and factors such as cable manufacturing stresses, environmental effects, and physical bends in the fiber all cause attenuation in an optical fiber.
- The amount of attenuation caused by an optical fiber is primarily determined by its length and the wavelength of the light traveling through the fiber.
- Fiber relative intensity noise (RIN) results from double Rayleigh scattering in optical fiber.
- Dispersion technically describes the spreading of the pulses of light as they travel down the optical fiber.
- Multimode dispersion describes the pulse broadening in multimode fibers caused by different modes traveling at different speeds through the fiber.
- Chromatic dispersion represents colors or wavelengths propagating at different speeds, even within the same mode.
- Waveguide dispersion occurs because optical energy travels in both the core and the cladding at slightly different speeds.
- Different wavelength dependencies of the refractive indices of the core and the cladding cause profile dispersion.
- Polarization mode dispersion (PMD) affects single-mode fibers; as two perpendicular polarization modes travel down the fiber, they may arrive at the end at different times.
- Dispersion management addresses the adverse effects of dispersion on transmission systems.
- Fiber nonlinearities raise the complexity of fiber optic system design; however, new fiber designs reduce the impact of the nonlinearities.
- Parameters such as fiber size, fiber material, attenuation, bandwidth-distance product, and the amount of power that can be coupled to the fiber, affect the choice of fiber for a given application.

Selected References and Additional Reading

Buck, John. *Fundamentals of Optical Fibers.* New York, NY: Wiley Interscience, 1995.

Calhoun, Dr. Hal. "Fiber Non-linearities and their Impact on Multi-gigabit Optical Transmission Systems." Paper presented at the National Fiber Optic Engineers Conference, June 18-22, 1995, Boston, MA, pp. 105-113.

Hecht, Jeff. *Understanding Fiber Optics.* 2nd edition. Indianapolis, IN: Sams Publishing, 1993.

Heismann, Fred. "Tutorial — Polarization Mode Dispersion: Fundamentals and Impact on Optical Communications Systems." ECOC '98, Vol. 2, September 20-24, 1998: 51-79.

Hentschel, Christian. *Fiber Optics Handbook.* 2nd edition. Germany: Hewlett-Packard Company, 1988.

Hluck, Laura L. "Optical Fibers for High Capacity Dense Wavelength-Division Multiplexed Systems." New Jersey: Corning Incorporated, 1998.

Just the Facts. New Jersey: Corning, Incorporated, 1992.

Lam, Jane and Liang Zhao. "Design Trade-offs for Arrayed Waveguide Grating DWDM Mux/Demux." NEED REST OF CITATION.

Lively, John. "Dealing with the Critical Problem of Chromatic Dispersion." *Lightwave Magazine.* September 1998: 77-79.

Sterling, Donald J. *Amp Technician's Guide to Fiber Optics,* 2nd Edition. New York: Delmar Publishers, 1993.

Hidenori Taga, Kaoru Imai, Noriyuki Takeda, Masatoshi Suzuki, Shu Yamamoto, and Shigeyuki Akiba. "100 WDM x 10 Gbit/s Long-distance Transmission Experiment Using a Dispersion Slope Compensator and Non-Soliton RZ Pulse. " *Optical Society of America, Trends in Optics and Photonics,* TOPS vol. XVI, July 21-23, 1997, pp. 382-385.

Tkach, Robert W. "Strategies for Coping with Fiber Nonlinearities in Lightwave Systems," Paper delivered at OFC '94, February 24, 1994.

Yanming Liu, Pam Dejneka, Laura Hluck, Curt Weinstein and Dan Harris. "Advanced Fiber Designs for High Capacity DWDM Systems," from *NFOEC Technical Proceedings, Volume 1.* Proceedings of the National Fiber Optic Engineers Conference, Orlando, FL, September 13-17, 1998.

FIBER OPTIC CABLES

Chapter 3 discussed the methods for manufacturing optical fiber and detailed many of its important characteristics. Another important characteristic is that bare optical fiber is quite delicate, making their handling difficult, especially in multifiber situations. To make handling optical fiber easier, it is first "buffered" or coated with a thin primary coating by the fiber manufacturer. It is then most often cabled. Fiber optic cables also function to protect the optical fiber from mechanical damage such as bends, breaks, or cuts both during and after installation. The exact structure and properties of a fiber optic cable are often directly related to the application for which the cable is constructed as well as the environment in which the cable will be installed.

FIBER OPTIC CABLE CONSTRUCTION

Although cable construction is usually application-specific, there are elements common to all types of fiber optic cables. The first element is the fiber housing. The fiber housing is either a loose-tube or a tight-buffer construction. Figure 4.1 shows the differences in these two types of construction in both cutaway and cross-section form.

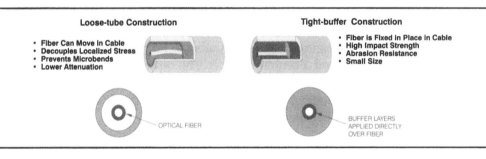

Figure 4.1: Fiber Optic Cable Construction

Loose-tube construction allows for lower attenuation but the tubes may permanently kink if the cable is bent beyond its limits, breaking the fiber. In loose-tube cable construction, the optical fibers are enclosed in plastic buffer tubes that are filled with a gel to impede water penetration. The buffer tubes are stranded around a dielectric or steel central member, which serves as an anti-buckling element. To save space, several fibers may be placed in the same tube before it is filled with the water-impeding gel. In high fiber counts, this is cost-effective, producing a lower cost per cabled fiber. Loose-tube cable construction is used mainly for long distance applications and permanent installations. A tight-buffer cable is smaller, more flexible, and more resistant to impact. In this cable design, the buffering material is in direct contact with the fiber. It is preferred when small size and high mechanical integrity are a main concern. In tight-buffer cable designs, the amount of material and effort used is more nearly proportional to the number of fibers; therefore, this design is more cost-effective in cables with lower fiber counts.

A key difference between loose-tube and tight-buffer cables is how they react to fluctuating temperatures. Since the fiber in a loose-tube cable is not mechanically coupled to the cable structure, it tends to do well in widely varying temperatures. The converse is true for most tight-buffer cable designs. The problem with the tight-buffer construction is that the materials used in the cable often have a higher thermal coefficient of expansion than the glass in the fiber itself. As the cable is heated, the fiber is stretched, and as the cable is cooled, the fiber is compressed. These stresses on the fiber cause the attenuation of the fiber to increase. This is usually referred to as cabling excess loss. This loss can range from a few hundredths of a dB per kilometer to several tenths of dB's per kilometer or more. Some very sophisticated fiber optic cable designs incorporate central strength members that have negative temperature coefficients of expansion so that the positive expansion caused by the outer jacket material is neutralized. Still, tight-buffer cable designs have some problems with temperature changes. Usually cold temperatures cause the most problems for fiber optic cables.

Ribbon cable is a variation of the loose-tube construction. A group of coated fibers is arranged so that the fibers are parallel to each other then coated with plastic to form a multifiber ribbon. Typically five to twelve fibers are encased in this manner. This set of fibers is then placed in one tube in the cable jacket and surrounded with gel, similar to other loose-tube designs. The simple structure of the ribbon cable makes it possible to splice all fibers of a ribbon simultaneously. This is a complex operation, but if done correctly, it can make splicing fast. The ribbon cable design allows for high packing density. However, mishandling during installation could cause uneven strain on different fibers, introducing the potential for uneven fiber losses.

The number of fibers used in the cable will affect the cable's current and future usability. Three parameters determine fiber count: the intended end-user applications, both present and future; the level of multiplexing and use of bridges/routers; and the physical topology of the network. Simple applications require a minimum of one fiber to establish basic communication; however, fiber count in a fiber optic cable ranges from the simplest single-fiber construction to complex multifiber cables. In the more complex, multifiber cable constructions, a common structure contains a number of buffered fibers or buffered fiber bundles loosely wound around a central member. The central member described above may be steel, coated steel, coated glass, a fiber glass rod, or simply filler. Small, indoor cables rarely include a central member, but central members are very common in outdoor cables and cables with high fiber counts.

Depending on the application, a variety of strength members can be designed into the cable construction. A common example is the use of Kevlar®, an extremely strong aramid yarn. This yarn can be applied longitudinally around each buffered fiber or used under the outer jacket of the cable. The use of Kevlar® allows the cable to withstand tension from 50-600 pounds for an extended period. It is common practice in cable installation to pull the cable through the duct work by the strength members to avoid putting stress on the fiber itself. In some cable designs, the strength members are integrated into the structure so that the entire cable may be pulled without concern over locating and attaching the strength members.

Figure 4.2: Two
Types of Cable Jackets

*(Photos courtesy of Optical Cable
Corporation.)*

The final component of a fiber optic cable is the cable jacket. The selection of jacket material requires consideration of such factors as mechanical properties, attenuation, environmental stress to be placed on the cable, and flammability. Variations in the outer jacket design allow the jacket to be interlocked to the cable core by filling the core's outer interstices. Strength members in the cable reinforce the jacket, preventing any movement of the jacket along the axis. Figure 4.2 shows two types of cable jackets, and Table 4.1 lists the properties of common cable jacket materials.

Jacket Material	Properties
Polyvinyl Chloride (PVC)	Normal mechanical protection. Many different grades of PVC offer flame retardancy and outdoor use. Also for indoor and general purpose applications.
Hypalon®	Can withstand extreme environments; flame retardant; good thermal stability; resistant to oxidation, ozone, and radiation.
Polyethylene (PE)	Used for telephone cables. Resistant to chemicals and moisture; low-cost. Polyethylene is flammable, so it is not used in electronic applications.
Thermoplastic Elastomer (TPE)	Low-cost; excellent mechanical and chemical properties.
Nylon	Used over single conductors to improve physical properties.
Kynar® (Polyvinylidene Fluoride)	Resistant to abrasions, cuts; thermally stable; resistant to most chemicals; low smoke emission; self-extinguishing. Used in highly flame retardant plenum cables.
Teflon® FEP	Zero smoke emission, even when exposed to direct flame. Suitable to temperatures of 200°C; chemically inert. Used in highly flame retardant plenum cables.
Tefzel®	Many of the same properties as Teflon; rated for 150°C; self-extinguishing.
Irradiated Cross-linked Polyolefin (XLPE)	Rated for 150°C; high resistance to environmental stress, cracking, cut-through, ozone, solvents, and soldering.
Zero Halogen Thermoplastic	Low toxicity makes it usable in any enclosed environment.
Kevlar, Hyplon, Tefzel, and Teflon are registered trademarks of E.I. Du Pont Nemours & Company. Kynar is a registered trademark of Pennwalt, Inc.	

Table 4.1: Properties of Cable Jacket Material

TYPES OF FIBER OPTIC CABLES

The simplex cable, illustrated in Figure 4.3, is round with a single fiber in the center. Duplex cables, having two fibers, may be circular, oval, or arranged zipcord fashion like an electrical cable. A zipcord cable is illustrated in Figure 4.4.

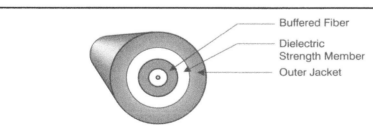

Buffered Fiber

Dielectric
Strength Member

Outer Jacket

Figure 4.3: Simplex Cable Construction

Figure 4.4: Duplex
Zipcord Cable
Construction

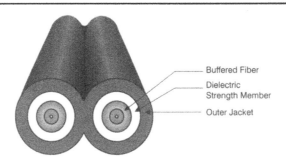

Figure 4.4: Duplex
Zipcord Cable
Construction

More complex cable assemblies fall into several categories including breakout (fanout) cables, composite cables, and hybrid cables. Breakout cables, also called fanout cables, are so called because the fibers are packaged in the cable as single-fiber or multifiber subcables. This allows the individual fibers to be accessed without the need for patch panels to terminate the multiple fibers. Figure 4.5 shows a cross-section of a typical breakout cable.

Figure 4.5: Multifiber
Breakout Cable
Construction

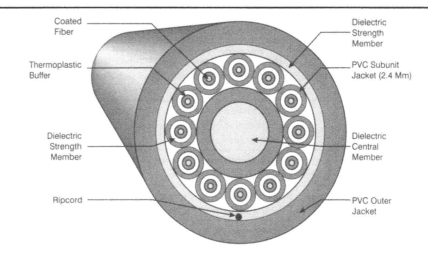

There is some disagreement in the cable industry as to the exact definitions of composite cables and hybrid cables. For the sake of this discussion, composite cables will be defined as cables mixing both single-mode and multimode optical fiber in a single cable assembly. The term hybrid cable will be used to describe a cable that mixes optical fiber with copper cable. Both composite cables and hybrid cables offer the advantage of time and cost savings. Multiple cables pulled separately would take longer to install. However, these cables are specialty items, custom-manufactured for a specific application, but the increase in demand for the convergence of video, audio, and data transmission make these types of cables a very attractive alternative to multiple separate cables. Hybrid cables can also carry electrical signals and are getting wide use in broadband CATV networks. Figure 4.6 shows a typical hybrid cable.

FIBER OPTIC CABLE VERSUS COPPER COAX CABLE

Copper coax cables and other conventional metal cables are the predecessors of fiber optic cables. Both fulfill the same basic function of signal transmission. Fiber optic cables resemble conventional cables and incorporate some of the same materials such as polyvinyl chloride (PVC) sheathing and polyethylene (PE) which is used for environmental protection, especially on buried and aerial cables. Optical fiber cable has many advantages

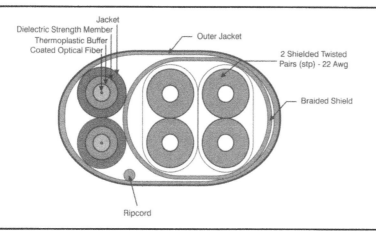

Figure 4.6: Hybrid Cable Construction

however. One major advantage is the nonconducting nature of optical fiber. While some fiber optic cables may contain conductive material such as a steel central member or steel outer sheath, most fiber optic cables are dielectric, meaning they contain no conductive materials. This gives the cable complete EMI/RFI immunity and reduces the impact of lightning strikes.

Figure 4.7 illustrates another advantage of optical fiber cable. The graph shows a comparison of the attenuation of a low-loss optical fiber and four popular types of copper coax cable. Note the very high levels of attenuation for the coax cable at higher frequencies.

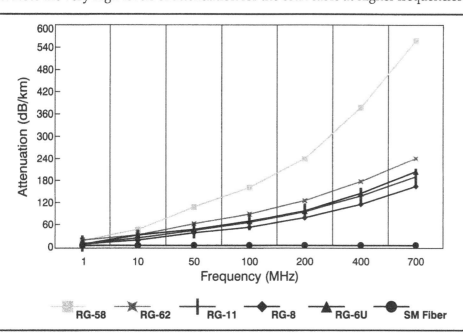

Figure 4.7: Comparison of Copper Coax versus Optical Fiber Cable

RG-6/U has an attenuation rate of 70 dB/km at a frequency of only 100 MHz. That means that only 0.00001% of the signal strength will reach the other end of a 1 km length of cable. Even the best coax shown, RG-8, has an attenuation rate of 45 dB/km at a frequency of 100 MHz. At a frequency of 40 MHz, RG-58/U has an attenuation rate of 100 dB/km. Only 0.00000001% of the signal strength would remain after 1 km of cable. By comparison, the single-mode fiber optic cable has an essentially flat 1 dB/km of attenuation, even at 500 MHz. This means that 79% of the signal strength will remain after a 1 km length of fiber.

The advantages of fiber at high frequencies, high data rates, and long distances are obvious. Even multimode fiber has only 8 dB of loss at 500 MHz, so 16% of the power remains after 1 km of fiber.

TENSILE STRENGTH

Fiber optic cables offer a higher mean tensile breaking strength than other types of cable. This value considers the pounds per square inch (psi) that can be applied before a given cable will snap. The depth of the microscopic flaws inherent on the surface determines the strength of an optical fiber. The theoretical mean tensile breaking strength of optical fiber is about 600,000 pounds per square inch or 600 kpsi, enough strength for a fiber with a diameter equivalent to one square inch to suspend fifty elephants in an elevator. Figure 4.8 shows the comparative tensile breaking strengths of various cable materials.

Figure 4.8: Mean Tensile Strength of Cable Materials

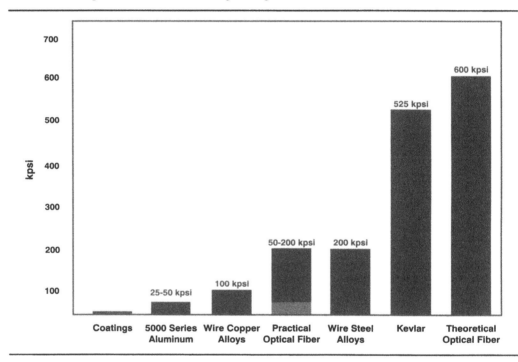

Usually, the theoretical fiber strength of 600 kpsi can only be achieved for very short lengths of fiber. Just as a chain is only as strong as its weakest link, a length of optical fiber is no stronger than the weakest point along its length. Microscopic cracks and flaws in the surface of the fiber limit the strength of the fiber to values typically less than 200 kpsi. All modern fibers are proof-tested at either 50 kpsi or 100 kpsi as they are manufactured. In special cases, fiber can be proof-tested to 200 kpsi or even higher for high-reliability applications.

The allowable bend radius is one important result of the fiber's proof-tested strength. A fiber with a proof-tested strength of 50 kpsi can have a minimum bend radius of 1.5". A fiber with a proof-tested strength of 100 kpsi can bend tighter to a 1" radius, and 200 kpsi fiber can bend as tight as a 0.75" radius.

Fiber with a higher proof-test will withstand more abuse during installation and throughout its lifetime. A fiber optic cable is generally assumed to have a 20 to 40 year life after it is installed. Removing the outer cable with mechanical strippers can degrade the fiber's life expectancy. Moisture and excessive stress are also enemies of optical fiber. Moisture can invade cracks in the outer surface of the fiber and can cause the cracks to propagate. Stress on a fiber can also cause those microcracks to propagate, ultimately causing the fiber to break. The proof-test level to which a fiber is subjected determines the maximum possible flaw size in the fiber. The maximum flaw size in 100 kpsi proof-tested fiber is about 1/3 the size of possible flaws in 50 kpsi proof-tested fiber. Flaw size greatly affects the lifetime of a fiber under given stress conditions. The 100 kpsi proof-tested fiber survives many times longer that the 50 kpsi proof-tested fiber.

SPECIFYING FIBER OPTIC CABLES

As mentioned, fiber optic cables are often custom manufactured from application to application. Cable environment is a critical parameter in determining the cable construction. There are many types of environments for fiber optic cables. Figure 4.9 illustrates some of these.

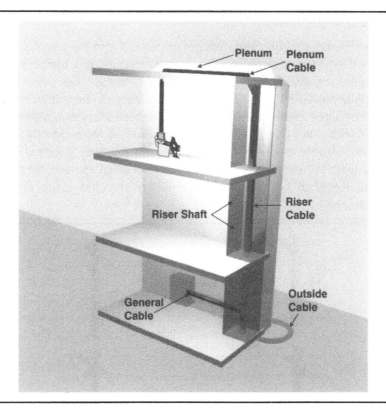

Figure 4.9: Cable Environments

The least strenuous cable environment can occur within devices such as computers, telephone switching systems, or distribution and splice organizers although high stresses can occur due to small bend radii. These cables are generally small and simple, and they tend to be low-cost. The device itself protects the cable from outside forces.

Intraoffice and intrabuilding cables are installed across a room, under a floor, between walls or above suspended ceilings, areas called plenum. These cables must meet fire and electrical safety standards before they are specified for installation in an office or an office building. The National Fire Protection Association publishes the National Electrical Code which defines safety considerations for using fiber optic cables within a building. There are three listings for cables: nonconducting optical fiber general purpose cable (OFN), nonconducting optical fiber riser cable (OFNR), and nonconducting optical fiber plenum cable (OFNP). Cable manufacturers must submit the cable design to an independent test laboratory such as Underwriter's Laboratories in order to obtain a listing for their cables. OFN cable is tested per UL 1581, a general test to determine the cable's resistance to generating smoke and spreading flames. OFNR cable is tested per UL 1666-1986, the riser flame test. This vertical flame test ensures that the fiber optic cable does not facilitate the spread of fire from floor to floor within the riser shaft of a building. OFNP cable is tested per UL 910, the plenum flame and smoke test. This horizontal flame test verifies that the cable will not spread flame or generate too much smoke within the plenum, ducts, or other spaces in a room or building.

Intraoffice and intrabuilding cables that meet these safety standards are often called plenum cables because that is usually where they are installed in the office or building. Intrabuilding cables are often constructed as breakout cables. This subcable design eliminates the need for patch panels in terminal closets. The subcables, which are color coded to simplify identification, allow the cable to be divided into individual fibers for distribution to separate end points in the office. Intraoffice cables tend to be of simpler construction, usually simplex or duplex cables.

Direct-burial cables are laid into deep trenches or plowed into the ground. This demanding environment requires extra protection against moisture and temperature extremes. Direct-burial cables usually incorporate an extremely strong outer jacket to protect the cable from damage caused by digging or chewing rodents. (In fact, military standards for fiber optic cables include rodent protection tests.) Moisture protection is often achieved by filling the cable with a gel that impedes the infiltration of water. Moisture is a concern because long-term exposure to moisture will degrade the fiber's optical characteristics. Extremely cold temperatures compound this problem when the moisture expands as it freezes, causing forces to be applied to the fiber that could cause microbends and increased optical loss.

Figure 4.10: Aerial Cable Installation

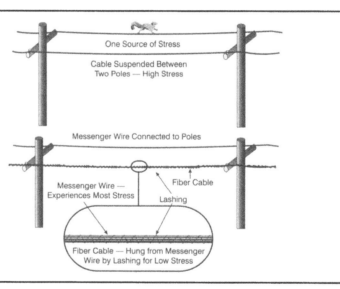

Fiber optic cables intended for aerial installation must also be constructed to handle environmental extremes. Aerial installation involves stringing the cable from utility poles as illustrated in Figure 4.10. Generally, these cables are all-dielectric i.e., they contain no metal. This prevents ground loops and provides the cable with lightning immunity. Suspension between two poles is the classic installation method. Internal strength members must be strong enough in this installation to prevent sagging that would put excess stress on the fibers. In other cases, the cable is wound or lashed to a parallel strength member. This supports the cable at more frequent intervals, reducing the stress along the length of the cable. Regardless of the actual installation method, all aerial cables have strength members and structures that isolate the fibers from the stress on the cable, and the outer jacket material offers UV protection from the sun in addition to protection from moisture and temperature extremes.

In situations where the cable must be used both within a building and outside from building to building, for example in a college campus network, indoor/outdoor cable may be specified. This cable type features the materials needed to meet fire and electrical safety standards encased in a removable layer of outdoor jacket material. Prior to the availability of indoor/outdoor cable, mixed environment cabling was handled in one of three ways: by using only tight-buffer cables (typically specified for indoor use), by using only loose-tube cables (typically specified for outdoor use), or by splicing combinations of tight-buffer cable inside the building to loose-tube cable outside the building. Loose-tube cables may be used in a building provided they are properly enclosed in metallic tubing per local building codes, and tight-buffer cables may be installed outside provided they are placed in a duct below the frost line. By splicing both types of cables together, each cable type is used in the most suitable environment. While the splice points will affect the optical link loss budget of the application and require additional labor for the splice, the cost differences between loose-tube cable and tight-buffer cable may make this an attractive solution in some applications when long outdoor runs are required.

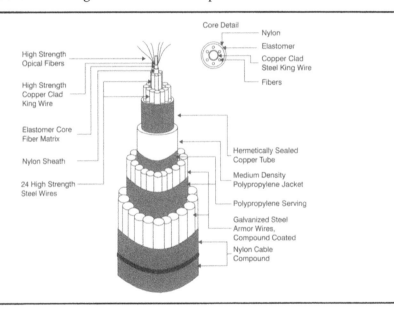

Figure 4.11:
Submarine Cable Construction

Another very demanding environment for fiber optic cable is the submarine environment where the cable is completely submerged in fresh or salt water. Submarine cables used for short distances are usually no more than ruggedized, waterproofed versions of direct-burial cable. Long-distance cables such as the one illustrated in Figure 4.11 are more complex. Damage can occur from boat anchors, trawlers, fishermen, and sharks as well as water pressure in applications where the cable is laid in very deep water. Three types of submarine cable systems impact the actual construction of the cable. Short systems, usually no more than 100 km, link islands with nearby continents or other islands. Systems longer than 1,000 km require the use of repeaters or optical amplifiers. Transoceanic cables, as the name suggests, run thousands of kilometers across oceans from continent to continent. These hybrid cables use copper cable to power the repeaters as well as optical fiber for the signal transmission. Higher data rates, exploitation of the 1550 nm window, and developing EDFA technology provide even greater distance capability for the optical signal and further reduce the need for repeaters in submarine systems.

While this chapter discusses many types of cables, new applications will bring new variety. The implementation of a global network has already created design challenges for fiber optic cable manufacturers. Still, the cable is but a conduit. Now it is time to turn the discussion to the principles behind the transmitters and receivers that use this conduit to carry information.

Chapter Summary

- Optical fiber is cabled to protect the fiber from damage during and after installation.
- Fiber cables can be either loose-tube or tight-buffer construction.
- Central members add rigidity to the cable.
- Strength members such as Kevlar® can be designed into the cable construction.
- The final component of a fiber optic cable is the cable jacket.
- The fiber count of a cable will affect the cable's present and future usability.
- Simplex cables have a single fiber in the center, while duplex cables contain two fibers.
- Complex cable assemblies include breakout, composite, and hybrid cables.
- Composite cables incorporate both single-mode and multimode optical fiber.
- Hybrid cables incorporate mixed optical fiber with copper cable.
- The least strenuous cable environment occurs within devices such as computers, telephone switching systems, or distribution and splice organizers.
- Direct-burial cables are laid into deep trenches or plowed into the ground and must be constructed to handle environmental extremes.
- Submarine cables being used for short distances are usually no more than ruggedized, waterproofed versions of direct-burial cable.
- Transoceanic cables run thousands of kilometers from continent to continent.

Selected References and Additional Reading

Just the Facts. New Jersey: Corning Incorporated, 1992.

Kachmar, Wayne M. *An Overview of Datacom Cable Installation: Installing the Fiber Network*. From the Proceedings of the Fiber Optic and Computer Networking Conference & Exhibition. October 24-26, 1990. Boston: World Trade Center, 1990.

Palladino, John R. *Fiber Optics: Communicating By Light*. Piscataway, NJ: Bellcore, 1990.

Sterling, Donald J. *Amp Technician's Guide to Fiber Optics*, 2nd Edition. New York: Delmar Publishers, 1993.

Universal Transport System Design Guide. Hickory, NC: Siecor Corporation, 1991.

LIGHT EMITTERS

5

Light emitters are a key element in any fiber optic system. These components convert the electrical signal into a corresponding light signal that can be injected into the fiber. The light emitter is important because it is often one of the most costly elements in the system, and its characteristics often strongly influence the final performance limits of a given fiber optic link.

There are two types of light emitters in widespread use in modern fiber optic systems: laser diodes (LD's) and light-emitting diodes (LED's). Laser diodes may be Fabry-Perot (the name refers to the mirrored surfaces on the ends), distributed feedback (DFB), or more recently vertical cavity surface-emitting laser (VCSEL) types while LED's are usually specified as surface-emitters or edge-emitters. These different classifications will be discussed in detail later in the chapter. All light emitters are complex semiconductors that convert an electrical current into light. The conversion process is fairly efficient in that it creates very little heat compared to the heat generated by incandescent lights. LED's and laser diodes are of interest for fiber optics because of five inherent characteristics:

- Overall small size
- High radiance (i.e., They emit a lot of light in a small area.)
- Small emitting area (The area is comparable to the dimensions of optical fiber cores.)
- Very long life (i.e. They offer high reliability.)
- Can be modulated (turned on and off) at high speeds

LED's and laser diodes are found in a variety of consumer electronics products. LED's are used as visible indicators in most electronics equipment, and laser diodes are most widely used in compact disk and DVD players. The LED's used in fiber optics differ from the more common indicator LED's in two ways:

1. The wavelength is generally in the near infrared range because the optical loss of fiber is the lowest at these wavelengths.
2. The LED emitting area is generally much smaller to allow the highest possible modulation bandwidth and improve the coupling efficiency with small core optical fibers.

THEORY OF OPERATION

Laser diodes and LED's operate on the same basic principle, the principle of the p-n semiconductor junction found in transistors and diodes. A positive-negative or p-n junction is comprised of a group IV element (Si, Ge, etc.) doped with a group III element (Al, Ga, In) and a group V element (P, As). (These element groups are based on the standard periodic table of elements, a copy of which can be found in most dictionaries.) The group III, IV, and V elements form a similar crystal lattice structure so the impurities added (group III and V) replace the group IV atoms on a one-to-one basis. The added group V atoms have one more valence electron than the group IV atoms, so they form an area with excess electrons called an n-type semiconductor. The added group III atoms have one less valence electron, forming an area called a p-type semiconductor. Although the doped semiconductor area has excess and deficient electrons, it is still overall electrically neutral.

When an n-type and a p-type semiconductor are placed together, the excess electrons from the n region move over into the p region to fill the holes left by the deficiency of electrons in the p-type material. The electrons stop moving into the p-type area when enough of them have built up to begin to repel any more electrons moving over (since electrons, all being negatively charged, repel each other). This buildup of charges creates a potential that prevents a current from flowing through the junction. Placing a potential across the p-n junction counteracts the internal potential enough to allow current to pass.

In direct semiconductors (semiconductors designed for optics), the electrons lose an amount of energy, called bandgap energy, corresponding to a property of the semiconductor material. This energy is released as photons that have a wavelength related to the bandgap energy by the formula:

$$\text{Eq. 5.1} \qquad\qquad E_g = \frac{hc}{\lambda} = \frac{1240 eV \cdot nm}{\lambda}$$

Where:

λ = Photon wavelength (nm).

E_g = Energy gap (eV).

The bandgap energy and wavelength of various semiconductors are listed in Table 5.1.

Table 5.1: Bandgap Energy & Wavelengths of Various Semiconductors

Material	Formula	Energy Gap	Wavelength
Gallium Phosphide	GaP	2.24 eV	550 nm
Aluminum Arsenide	AlAs	2.09 eV	590 nm
Gallium Arsenide	GaAs	1.42 eV	870 nm
Indium Phosphide	InP	1.33 eV	930 nm
Aluminum Gallium Arsenide	AlGaAs	1.42-1.61 eV	770-870 nm
Indium Gallium Arsenide Phosphide	InGaAsP	0.74-1.13 eV	1100-1670 nm

LIGHT EMITTER PERFORMANCE CHARACTERISTICS

Several key characteristics of LED's and lasers determine their usefulness in a given application. These are:

Peak Wavelength: This is the wavelength at which the source emits the most power. It should be matched to the wavelengths that are transmitted with the least attenuation through optical fiber. The most common peak wavelengths are near 780 nm, 850 nm, 1310 nm, 1550 nm, and 1625 nm.

Spectral Width: Ideally, all the light emitted from an LED or a laser would be at the peak wavelength, but in practice, the light is emitted in a range of wavelengths centered at the peak wavelength. This range is called the spectral width of the source.

Emission Pattern: The pattern of emitted light affects the amount of light that can be coupled into the optical fiber. Ideally, the size of the emitting region should be similar to the diameter of the fiber core.

Power: The best results are usually achieved by coupling as much of a source's power into the fiber as possible. The key requirement is that the output power of the source be strong enough to provide sufficient power to the detector at the receiving end, considering fiber attenuation, coupling losses and other system constraints. In general, lasers are more powerful than LED's.

Speed: A source should turn on and off fast enough to meet the bandwidth limits of the system. The speed is given according to a source's rise or fall time, the time required to go from 10% to 90% of peak power. Lasers have faster rise and fall times than LED's.

SPECTRAL CHARACTERISTICS

Spectral characteristics in LED's and lasers are illustrated in Figure 5.1 shows.

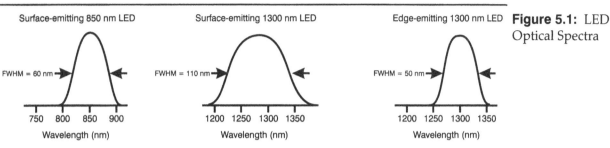

Figure 5.1: LED Optical Spectra

An 850 nm surface-emitting LED has a FWHM of 60 nm. FWHM stands for full width half maximum, a measure of the width of the optical spectrum taken at the point where the intensity falls to half of the maximum value. A surface-emitting 1300 nm LED has an even wider spectrum with a typical FWHM of 110 nm. This wider spectrum causes increased dispersion and also poses difficulties in WDM systems. A 1550 nm surface-emitting LED would even be wider. The last figure shows an edge-emitting 1300 nm LED. It has a much more compact spectrum with a typical FWHM of about 50 nm.

Figure 5.2 shows similar information for laser diodes. There are two different types of lasers shown. First, a 1300 nm Fabry-Perot laser is shown. The spectrum consists of nine discrete lines. This would properly be called a multimode laser, not referring to multimode fiber, but to the fact that the laser emits light at a number of discrete frequencies. Unlike a helium-neon (HeNe) laser which emits a single line at 633 nm, Fabry-Perot lasers emit at multiple, closely spaced wavelengths simultaneously. It can be said that light from diode lasers is less coherent than the light from a HeNe laser. This wider spectrum does cause some additional dispersion in the fiber, but it minimizes a nasty problem that may occur when using multimode fiber called modal noise which occurs whenever the optical power propagates through mode-selective devices, such as connectors and splices.

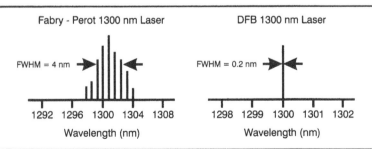

Figure 5.2: Laser Optical Spectra

A HeNe laser spot has a sparkle in its appearance, known as modal noise, which is caused by constructive and destructive interference of the light producing dark and light spots. Highly coherent diode lasers produce a similar speckle pattern inside an optical fiber. Connectors and discontinuities in the fiber disturb this pattern, resulting in a sharp increase in system noise. Multimode lasers are a better choice when used with multimode fiber since they are less coherent and produce a lower contrast speckle pattern. The DFB laser is almost perfectly coherent, emitting a single frequency, like the HeNe laser, making it prone to modal noise.

LIGHT-EMITTING DIODES

LED's are made of several layers of p-type and n-type semiconductors. A p-n junction generates the photons, and several p-p and n-n junctions direct the photons to create a focused emission of light. The p-p and n-n junctions direct the light by providing energy barriers and changes in the index of refraction. There are two main types of LED's currently being used, surface-emitters and edge-emitters.

Figure 5.3: Typical Packaged LED's

(Photo courtesy of PD-LD, Inc.)

Surface-emitters are made of layers of semiconducting material that emit light in a 180° arc. They are relatively inexpensive and very reliable, but the emission pattern limits the coupling efficiency with the fiber, and therefore, the power that can be transmitted. Surface-emitters are the most economical of the two types of LED's, but they have low output power and are generally slower devices.

Figure 5.4: Surface-emitting LED

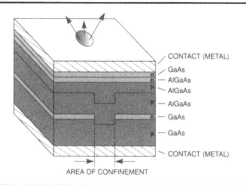

Edge-emitters are designed to confine the light to a narrow path directed out the side of the emitter. This focusing of the light means more power is emitted, and more power can be coupled to the fiber because the path is comparable to the size of the optical fiber core. Edge-emitters can provide high optical power levels (even into smaller core fibers) and are generally faster devices than surface-emitting LED's.

Figure 5.5: Edge-emitting LED

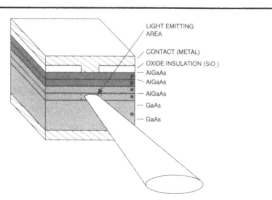

LED's can survive and operate in extreme temperature ranges (up to full military specifications, -55°C to +125°C), although their optical output power can drift considerably as temperature varies. In general, shorter wavelength LED's are less prone to drift over temperature. Also, surface-emitting LED's are almost always more stable over temperature than edge-emitting types. An 850 nm LED may only drift -0.03 dB/°C, while a 1310 nm LED may drift three to five times as much. Always, the LED's optical power drops as the temperature increases.

Temperature also affects the peak emission wavelength. Most LED's exhibit a 0.3 nm/°C to 0.6 nm/°C drift in the peak emission wavelength as temperature varies. This can be important in multi-wavelength systems where there is a possibility of crosstalk, the interference of one channel on another, if the LED wavelength drifts.

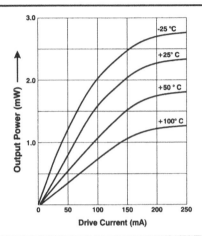

Figure 5.6: Typical LED Behavior versus Temperature

Of the two light source types, LED's are the most widely used for short system fiber optic applications. In general, LED's tend to cost less than laser diodes, although the cost advantage has diminished some, so they find wider application. A common trait of all LED types is that they turn on much faster than they turn off. This is due to the long carrier recombination lifetimes in LED's. Often, this carrier recombination will be in the tens of nanoseconds range. Most LED driver circuits employ special tricks to overcome this, such as applying quick reverse bias to the LED at turn-off to sweep the carriers out of the active region. Even with these circuit tricks, LED's are limited to lower speed applications than laser diodes. Most commercial LED's have a top-end bandwidth of 100-200 MHz. A few specialty LED's are available with bandwidths as high as 600 MHz, according to manufacturer's claims, but still, LED's are only widely used for data rates below 270 Mb/s. Part of the reason is that the wider spectral bandwidth of LED's causes more dispersion in the optical fiber. This increased dispersion severely limits optical signal's maximum transmission distance.

The availability of low-cost CD and DVD lasers has also limited the use of LED's in high data rate applications. These lasers are manufactured in very high volumes (over one million a month) making them very economical. While these lasers were not specifically developed for data applications, they often do very well at data rates of one gigabit per second or more.

LED Driver Circuits

LED optical output is approximately proportional to drive current. Other factors, such as temperature, also affect the optical output. Figure 5.7 shows the typical behavior of an LED. Two curves are shown. The top curve represents a 0.1% duty cycle with the peak current as shown on the horizontal axis. The bottom curve shows the output with 100% duty cycle. The drop is primarily due to the heating of the LED chip. Most LED's have light versus current curves that droop or fall below a linear curve. A few, such as super-radiant LED's (near lasers) can have curves that bow upward rather than downward.

Figure 5.7: Optical Output versus Current in an InGaAsP Light-emitting Diode

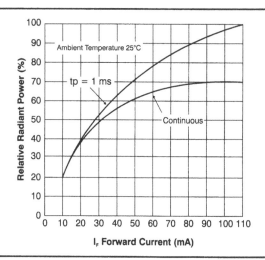

LED's are usually driven with either a digital signal or an analog signal. When the drive signal is digital, there is no concern about LED linearity. The LED is either on or off. There are special problems that need to be addressed when designing an LED driver. The key concern is driving the LED so that the maximum speed is achieved. Figures 5.8a, 5.8b and 5.8c show three popular digital LED driver circuits.

Figure 5.8: LED Driver Circuits

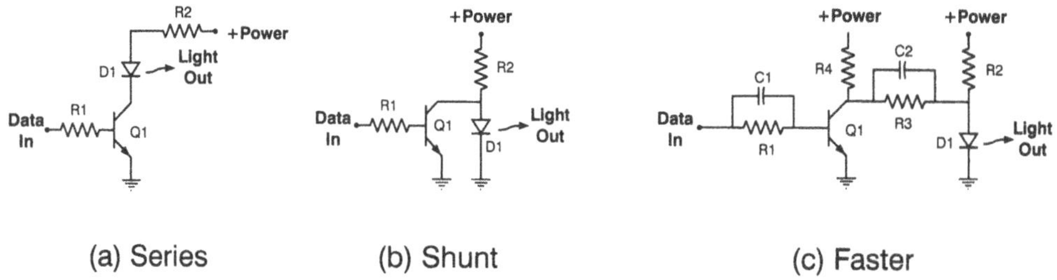

(a) Series   (b) Shunt   (c) Faster

The first circuit, shown in Figure 5.8a, is a simple series driver circuit. The input voltage is applied to the base of transistor Q1 through resistor R1. The transistor will either be off or on. With Q1 in the off condition, no current will flow through the LED, and no light will be emitted. When transistor Q1 is on, the cathode (bottom) of the LED will be pulled low. Transistor Q1 will pull its collector down to about 0.25 Volts. The current is equal to the voltage across resistor R2 divided by the resistance of R2. The voltage across R2 is equal to the power supply voltage less the LED forward voltage drop and the saturation voltage of the drive transistor. This circuit offers the key advantage of a low average power supply current. If one defines the peak LED drive current as I_{LEDmax} and assumes that the LED duty cycle is 50%, then the average power supply current is only $I_{LEDmax}/2$. Further, the power dissipated is $(I_{LEDmax}/2) \cdot V_{SUPPLY}$ where V_{SUPPLY} is the power supply voltage. The power dissipated by the individual components, the LED, transistor and resistor R1, is equal to the voltage drop across each component multiplied by $(I_{LEDmax}/2)$. The key disadvantage of the circuit shown in Figure 5.8a is low speed. This type of driver circuit is rarely used at data rates above 30-50 Mb/s. In general, there are two ways to design an LED drive circuit for low power dissipation. The first is to use a high-efficiency LED and reduce I_{LEDmax} to the lowest possible value. The second is to reduce the duty cycle of the LED to a low value. Usually larger gains can be made with the second method.

The second LED driver circuit, shown in Figure 5.8b, offers much higher speed capability. It uses transistor Q1 to quickly discharge the LED to turn it off. This circuit, known as a shunt drive circuit, will drive the LED several times faster than the series drive circuit shown in Figure 5.8a. The shunt drive circuit gives greater drive symmetry. In the shunt driver circuit in Figure 5.8b, resistor R2 provides a positive current to turn on the LED. Typically, R2 would be in the 40 Ohm range. This makes the turn-on current about 100 mA peak. Transistor Q1 provides the turn-off current. When saturated, transistor Q1 will have an impedance of a few Ohms. This provides a much larger discharging current allowing the LED to turn off quickly. Power dissipation typically more than doubles in a shunt driver circuit compared to the series driver. In fact, the circuit draws more current and power when the LED is off than when the LED is on! The exact power dissipation can be computed by first analyzing the off and on state currents and then combining the two values using information about the operating duty cycle.

Figure 5.8c shows a variation on the shunt driver shown in Figure 5.8b. Two additional resistors and two capacitors have been added to the basic circuit. These additional components further improve the operating speed. Capacitor C1 improves the turn-on and turn-off characteristics of transistor Q1 itself. Care should be taken to keep the value of C1 down to prevent the transistor base from being overdriven and damaged. The additional components, resistors R3 and R4, and capacitor C2 provide overdrive when the LED is turned on and underdrive when the transistor is turned off. The overdrive and underdrive accelerates the LED transitions. Typically, the RC time constant of R3 and C2 approximately equals the rise or fall time of the LED itself when driven with a square wave. All of these tricks together can increase the operating speed of the LED and driver circuit to about 270 Mb/s. There have been numerous laboratory tests and prototype circuits that have achieved rates to 500-1000 Mb/s, but none of these have made it into mass production. Typically, these levels of performance require a great deal of custom tweaking on each part to achieve the high data rates.

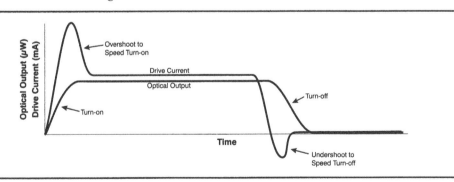

Figure 5.9: Response of an LED to a Digital Modulation Signal

Figure 5.9 shows the response of an LED to a digital modulation signal. The electrical signal shown is the type generated by more sophisticated LED driver circuits such as that shown in Figure 5.8c. Starting at time zero, we first see the digital signal go to a logic level 1. The most remarkable part of this event is the strong overshoot seen on the electrical drive signal. This overshoot may be two times the steady state logic 1 drive current, and it accelerates the turn-on time or rise time of the LED. Even so, we see that the optical output lags behind the electrical signal.

Typical values for very high-performance LED's and driver circuits would be 0.7 ns rise time of the electrical signal and 1.5 ns optical rise time. Later, when the digital signal goes back to a logic *0*, we see the same process repeated. The electrical signal has a strong undershoot component which acts to accelerate the turn-off of the LED. The undershoot serves to reverse bias the LED, sweeping out the carriers. Even so, the turn-off time of most LED's is always slower than the turn-on time. Typical values for turn-off times are 0.7 ns for the electrical signal and 2.5 ns for the optical signal. Note that while in a logic *0* state, the drive current does not quite go to zero. It is common to provide a small amount of pre-bias current, typically a few percent of the peak drive current, to keep the LED forward biased and improve dynamic response.

LASER DIODES

There are two main types of laser diode structures, Fabry-Perot (FP) and distributed feedback (DFB). DFB lasers offer the highest performance levels and also the highest cost of the two types. They emit nearly monochromatic light (i.e. they emit a very pure single color of light) while FP lasers emit light at a number of discrete wavelengths. DFB lasers tend to be used for the highest speed digital applications and for most analog applications because of their faster speed, lower noise, and superior linearity.

Fabry-Perot lasers further break down into buried hetero (BH) and multi-quantum well (MQW) types. BH and related styles ruled for many years, but now MQW types dominate. MQW lasers offer significant advantages over all former types of Fabry-Perot lasers. They offer lower threshold current, higher slope efficiency, lower noise, better linearity, and much greater stability over temperature. As a bonus, the performance margins of MQW lasers are so great, laser manufacturers get better yields, so laser cost is reduced. One disadvantage of MQW lasers is their tendency to be more susceptible to backreflections. MQW lasers also perform poorly as detectors, making them ill-suited for ping-pong applications (described at the end of this chapter).

Most laser diodes used for fiber optic communications incorporate a rear facet photodiode to provide a real-time means of monitoring the output of the laser. This is necessary because the threshold current of the laser changes with temperature as does slope efficiency. (Figure 5.19, later in this chapter, shows these effects.) Vertical cavity surface-emitting lasers (VCSEL's), discussed later, are one of the exceptions. They have very stable characteristics over time and often omit monitor photodiodes.

Figure 5.10: Laser Construction

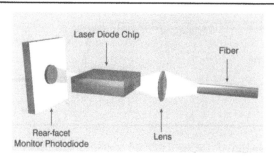

Figure 5.10 shows the typical optical construction of a laser diode. The laser diode chip emits light in two directions. The light from one end of the laser chip is focused onto the fiber and provides the useful output. The light from the other end falls on a large area photodiode. Usually, this photodiode is mounted some distance from the laser chip and is angled to reduce backreflections into the laser cavity.

Fabry-Perot lasers are the most economical, but they are generally noisy, slower devices. DFB lasers are quieter devices (e.g., exhibit a high signal-to-noise ratio) have narrower spectral width, and are usually faster devices. Laser noise is expressed as relative intensity noise (RIN) in units of dB/Hz. Typical RIN values for good Fabry-Perot devices are -125 to -130 dB/Hz while a good DFB laser can have RIN values below -155 dB/Hz. DFB lasers designed for continuous wave (CW) operation can have RIN values below -170 dB/Hz. To convert RIN values into a resultant signal-to-noise ratio, take the absolute RIN value and subtract $10\log_{10}$ (BW) where BW is the bandwidth of interest in Hertz. For instance, if the laser has a RIN of -130 dB/Hz and the system has a bandwidth of 1 MHz, the signal-to-noise ratio of the laser's optical signal would be 70 dB.

VCSEL's

The first commercialized laser chips, usually called edge-emitters, output laser light from the edges, and the laser cavity ran horizontally along the laser's length. A new laser structure being used to produce low-cost lasers, the vertical cavity surface emitting laser or VCSEL, emits laser light, as the name suggests, vertically from its surface and has a vertical laser cavity.

The VCSEL's principles of operation closely resemble those of conventional edge-emitting semiconductor lasers. At the heart of the VCSEL an electrically pumped gain region, also called the active region, emits light. Layers of varying semiconductor materials above and below the gain region create mirrors. Each mirror reflects a narrow range of wavelengths back into the cavity causing light emission at a single wavelength. Figure 5.11 illustrates this structure

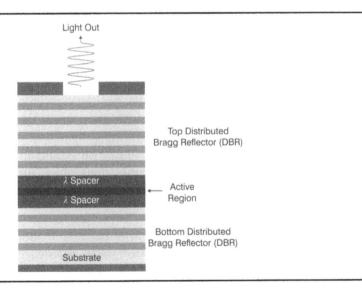

Figure 5.11: Basic VCSEL Structure

VCSEL's are typically multi-quantum well devices with lasing occurring in layers only 20-30 atoms thick. Bragg reflectors with as many as 120 mirror layers form the laser reflectors. Because of the small size of the VCSEL's and the high efficiency of the mirrors, the threshold current can be very low, below 1 mA. VCSEL's also exhibit very high slope efficiency. Early devices for use with multimode fiber have an optical output of several milliwatts at a drive current of 12 mA and a typical slope efficiency of 200 μW/mA.

The transfer function of a VCSEL (light out versus drive current) allows stability over a wide temperature range, a feature unique to these laser diodes. Many manufacturers recommend that their VCSEL's operate at a current of 12 mA and even omit the customary monitor photodiode. VCSEL's have low threshold, low operating current, and high slope efficiency because they use more efficient mirrors than earlier laser types. Earlier laser types generally used cleaved ends of the laser chip to form reflectors; however, these reflectors rarely reflected more than 35% of the light. VCSEL's use distributed Bragg gratings (DBR's) formed of a hundred or more alternating layers of material that efficiently reflects light at a specific wavelength. A more efficient reflector allows the light intensity to build up at lower drive levels.

VCSEL's have several advantages over edge-emitting lasers. Manufacturer's create edge emitters on a substrate, and they must be cut out of the substrate before packaging them for testing. This expensive process can be more costly than the laser diode itself, and if the laser then fails tests, the effort has been wasted. Because VCSEL's emit light vertically from their surface, they can be tested while they are still on the "wafer" of substrate material (illustrated in Figure 5.12), saving time and money. Additionally, VCSEL's have a narrower, more circular light beam compared to edge emitters, making it easier to get the light into an optical fiber. The lower output power and lack of longer wavelengths of today's VCSEL's limits their usefulness in some applications. However, their cost advantage makes them an attractive option for future applications.

Figure 5.12: VCSEL Wafer

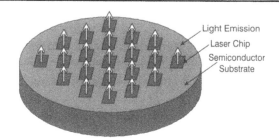

Because of the way VCSEL's are manufactured, they are ideal for applications that require arrays of devices. For example, one could create a linear array of lasers for use with ribbon optical fiber. This allows multiple parallel data links to be readily formed. This is useful where 8-bit bytes of data must be moved from point to point. Conventional solutions would involve serializing the bytes, transmitting them over a single fiber, and then reconstructing the parallel bytes at the receive end. With VCSEL arrays and ribbon fiber, each of the eight bits is transmitted with its own laser and fiber. This simplifies the electronics at each end and allows much higher data rates to be moved.

CHIRP

Variations in the wavelength of a laser, in response to modulation, causes the phenomenon called laser chirp. Directly modulated lasers experience a modulation current that causes the laser's wavelength to shift proportionally, creating laser chirp. While one usually considers that the modulation current applied to a laser is AM modulating the laser output, that same current is also FM modulating the laser output. The frequency (color) of the emitted laser light actually changes in response to the modulation current. For a typical DFB laser, the frequency shift can be about 100 MHz per mA of drive current. Figure 5.13 shows this phenomenon.

While the laser may output a very narrow temporal pulse, the pulse contains a range of optical wavelengths. The spread of optical wavelengths interacts with the dispersion in the fiber. The end result of chirp leads to pulse broadening as it interacts with the fiber. At high data rates, the impact of pulse broadening becomes more critical because of the inherently shorter unit intervals (or bit-times) at these rates. In DWDM systems, the

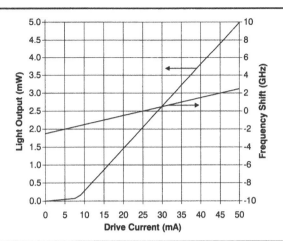

Figure 5.13: Laser Chirp: DFB – Direct Modulation

narrower channel spacing at high speeds also increases the effects of channel crosstalk. External modulation devices prevent laser chirp while simple modulation techniques may minimize the phenomenon.

Whether minimizing or preventing laser chirp, measuring its effects in high-speed DWDM systems can be critically important in determining the suitability of the laser in a data transmission system. Conventional applications must be designed to avoid laser chirp. It can, however, create a super wideband FM modulator that would allow, for instance, the conversion of the complete CATV spectrum (40 MHz to 860 MHz) to a 5 GHz wide FM signal. Figure 5.14 below shows the basic technique. One laser is direct modulated with the wideband AM signal. A second laser is operated at a constant current. The output of the two lasers are mixed together using a fiber optic coupler.

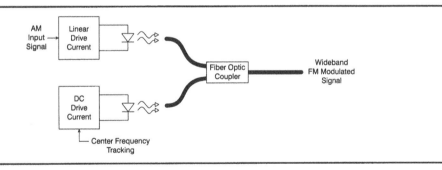

Figure 5.14: Wideband FM Modulator

BACKREFLECTION

All lasers are susceptible to backreflections. Backreflections disturb the standing-wave oscillation in the laser cavity, increasing the effective noise floor of the laser. A strong backreflection causes certain lasers to become wildly unstable and completely unusable in some applications. Backreflection can also generate nonlinearities in the laser response which are often described as kinks. Most analog applications and some digital applications cannot tolerate these degradations.

The importance of controlling backreflections depends on the type of information being sent and the particular laser. Some lasers are very susceptible to backreflections due to the design of the laser chip itself. Most often, the determining factor is how tightly the fiber is coupled to the laser chip. In a low-power laser potentially only 5-10% of the laser power is coupled into the fiber. This means that only 5-10% of the backreflection would be coupled into the laser cavity, making the laser relatively immune to backreflections. On the other hand, a high-power laser may have 50-70% of the laser chip output coupled to the fiber. This also means that 50-70% of the backreflection will be coupled back into the laser cavity. This makes high-power lasers more susceptible to backreflections.

One strategy to reduce backreflections places an optical isolator at the laser output. A Faraday rotator, illustrated in Figure 5.15 and based on the Faraday effect, is one example of an optical isolator.

Figure 5.15: Isolator Based on Faraday Rotator

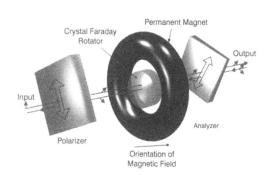

Before entering the Faraday rotator, which is usually an yttrium-iron-garnet (YIG) material, the light beam passes through a polarizer which is oriented parallel to the incoming state of polarization. The Faraday rotator then rotates the polarization by 45°. At the output, the beam passes an analyzer which is oriented at an angle of 45° relative to the first polarizer. Of all possible reflected beams, only those with a 45° orientation of the polarization are allowed to pass backwards. The polarization of the reflected beam is rotated by another 45° which results in a total rotation of 90°. This way, the reflected beam is blocked by the polarizer, reducing backreflections by 20 to 45 dB. In order not to disturb the proper function of the isolator, all of its surfaces should be antireflection-coated.

The isolator design in Figure 5.15 works with polarized light. In newer designs, the input and output polarizers are replaced with birefringent crystals which eliminate the sensitivity to the polarization of the light.

Unfortunately, optical isolators do not substitute for properly polished, low-backreflection connectors. The amount of rejection offered by an optical isolator will improve problems caused by backreflections but often will not eliminate them. The effects of these backreflections can disrupt a fiber optic transmission system. Figure 5.16 shows a laser waveform with no backreflection, and Figure 5.18 shows a laser waveform with a strong backreflection. First, the waveform in Figure 5.16 needs some explanation.

Figure 5.16: Laser Optical Output With No Backreflection

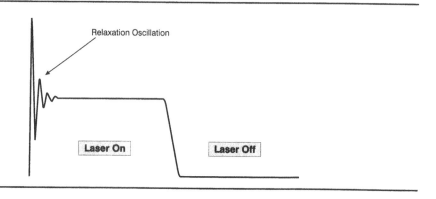

As seen in the figure, the rising edge is followed by a damped oscillation. This overshoot and the subsequent oscillation is called the relaxation oscillation. Most lasers exhibit this phenomenon. It can be understood by looking at the frequency versus amplitude response of the laser as illustrated in Figure 5.17.

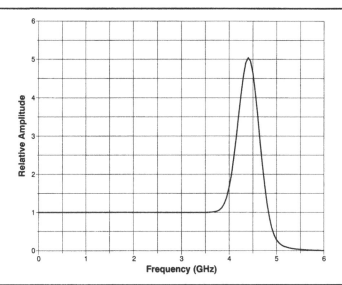

Figure 5.17:
Frequency versus Amplitude Response of a Laser

The laser exhibits a resonance (high gain point) near 4.4 GHz in this case. This will be the approximate frequency of the relaxation oscillation. The frequency of the resonance peak is a factor that limits the maximum data rate a given laser can transmit. When the maximum frequency component of the data stream gets close to the laser resonance frequency, performance degrades quickly. The frequency of the resonance and the magnitude of the overshoot depend on the drive levels applied to the laser. Overshoot is generally most severe when the laser is turned completely off and then back on. This condition is avoided in most practical data links by keeping the laser always above the threshold.

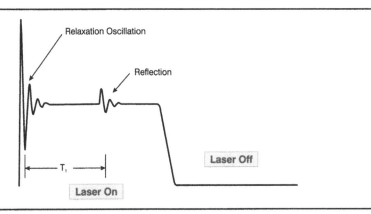

Figure 5.18: Laser Optical Output With Backreflection

The backreflection illustrated in Figure 5.18 shows the same characteristics as the initial relaxation oscillation. The time, T_1, can be precisely measured to determine the distance to the reflecting point. Often, the frequency has to be greatly reduced to observe such reflections. One fallacy is that only close reflections matter. Closer reflections are often a bit stronger, but at 1310 nm or 1550 nm, fiber loss is so low that reflections from a distance of several kilometers can be significant.

Backreflections can be observed by monitoring the photodiode servo-loop for disturbances. To do this, place the end of the laser pigtail in glycerin, which will eliminate virtually all backreflections. Then note the output of the servo loop at that time. Afterwards, connect the laser pigtail to the system. Any significant perturbations noted are backreflections of sufficient amplitude to disturb the standing wave in the laser cavity. This directly observes laser backreflections. A less direct test for laser backreflections involves testing at frequencies where backreflections will occur at an exact multiple of the bit time. Basically, this procedure calculates the round trip transit time to the potential reflection interfaces in the system. It is generally easiest to measure the spacing between the high interference points when using this method. Often, the laser pigtail is made very short so that the first reflection occurs at a frequency higher than any frequency being transmitted by the system. However, longer fiber segments in the system will yield a low fundamental interference frequency and harmonics that will clutter the spectrum. Designing laser-based systems for low backreflections remains the only practical strategy.

TEMPERATURE EFFECTS ON LASERS

Lasers can survive wide temperature ranges (up to full industrial specifications, -40°C to 85°C). Temperature affects the peak emission wavelength as well as the threshold current and the slope efficiency of the laser. Most uncooled FP lasers exhibit a 0.3-0.6 nm/°C drift in the peak emission wavelength as temperature varies. Uncooled DFB lasers have about an order of magnitude less drift. Cooled lasers have an extremely stable peak wavelength. This is crucial for dense wavelength-division multiplexing (DWDM) applications which use very closely spaced optical wavelengths. Optical output power can be relatively stable over a wide temperature range because of the servo-loop that typically stabilizes the output power based on the internal monitor photodiode. The internal photodiode monitors the rear facet of the laser chip while the front facet output is coupled to the optical fiber. If the front and rear facet outputs are perfectly correlated and the mechanical assembly is unaffected by temperature, then the optical output power would be perfectly stable as temperature varied. Neither is true, however, so the output power is affected by temperature. This effect is called tracking error. It is generally less than a few dB's over the temperature range, although its direction and magnitude vary widely from part to part.

Figure 5.19: Laser Optical Power Output versus Forward Current

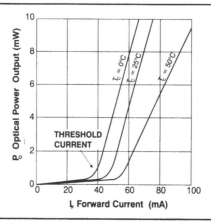

Laser optical output is approximately proportional to the drive current above the threshold current. Below the threshold, the output is from the LED action of the device. Above the threshold, the output dramatically increases as the laser gain increases. Figure 5.19 shows the typical behavior of a laser diode. As operating temperature changes, several effects occur. First, the threshold current changes. The threshold current is always lower at lower temperatures and vice versa. Slope efficiency, the second important parameter, also changes. The slope efficiency is the number of milliwatts or microwatts of light output per milliampere of increased drive current above threshold. Most lasers show a drop in slope efficiency as temperature increases. Figure 5.19 shows that the 50°C curve is not as steep as the curve at the lower temperatures which translates to lower slope efficiency.

PELTIER COOLERS

A decade ago, most lasers used in fiber optic links were thermoelectrically (TE)-cooled devices. Peltier coolers were used to stabilize the temperature of lasers. These devices are specialized semiconductors that act like miniature heat pumps, similar in function to those found in many homes. They are built from stacks of p and n elements made from doped bismuth telluride ($BiTe_2$). Peltier coolers pump heat from one end to the other when a current is passed through the device. The process is fully reversible. If one reverses the direction of current flow through the elements, the direction of heat transfer is reversed as well.

Small Peltier coolers were used to heat or cool the laser chip so that it could be maintained at a constant temperature. The laser package also generally included a thermistor so that the laser temperature could be accurately measured along with the customary rear-facet monitor photodiode. Cooled lasers were prevalent in the early days because long-term laser stability and life could only be achieved by maintaining the laser at tightly controlled temperatures. Today, the most reliable lasers are uncooled. Cooled lasers are somewhat less reliable than uncooled lasers because of the fragile nature of the Peltier cooler elements, especially at high ambient temperatures. Curiously, cooled lasers cannot cope with as wide a temperature range as uncooled lasers. This is true for two reasons. First, single-stage Peltier coolers, the type most often used with lasers, can only maintain about a 40°C temperature delta. Second, the Peltier cooler elements are usually assembled using low-melting-point solder which limits their top end temperature range.

Today cooled lasers are used for two main purposes. Used in conjunction with very high-power lasers, they remove heat from the laser chip, and they assure wavelength stability of the laser. It was mentioned earlier that the center wavelength of a laser (or LED) varies by about 0.03-0.6 nm/°C. This wavelength variation can cause three problems. First, it can affect the bandwidth that can be achieved over very long lengths of fiber. Long-haul systems often try to operate as close to the zero-dispersion wavelength of the fiber as possible. A drift of a few nanometers can cause the fiber's bandwidth to drop by a factor of ten or more. Second, it can affect the attenuation of the fiber over very long distances. This is usually only a problem at the 1550 nm wavelength. Last, laser center wavelength drift can cause unacceptable levels of crosstalk in wavelength-division multiplexed (WDM) systems with catastrophic results. The recent DWDM explosion has created a new market for cooled lasers. In this case, the center wavelength must be stable to roughly 0.01 nm.

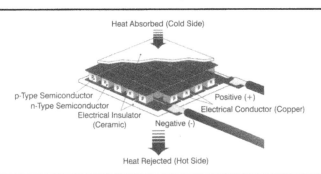

Heat Absorbed (Cold Side)

p-Type Semiconductor
n-Type Semiconductor
Electrical Insulator
(Ceramic)

Positive (+)
Electrical Conductor (Copper)
Negative (-)

Heat Rejected (Hot Side)

Figure 5.20: Cross-section of a Single-stage, Multi-element Peltier Cooler

(Illustration courtesy of Melcor Corp.)

Figure 5.20 shows a drawing of a single-stage, multi-element Peltier cooler. The heat flows in the direction shown as long as the electrical polarity is as shown. If the electrical polarity reverses, the direction of heat flow will reverse as well. The amount of heat pumped from one surface to the other depends on the amount of current flowing through the cooler. Typical laser and detector Peltier coolers require ±1 Amp at 1 to 2 Volts maximum.

With Peltier coolers, good thermal design is essential. The design must provide for adequate heat sinking available on the non-laser side of the device, or the Peltier cooler will overheat and destroy itself. Figure 5.21 shows a Peltier cooler mounted in a 14-pin laser package.

Figure 5.21: A Peltier Cooler Mounted in a 14-pin Laser Package

(Photo courtesy of Melcor Corp.)

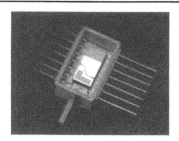

HIGH POWER AND CONTINUOUS WAVE (CW) LASERS

In recent years there has been a strong move to increase the optical power output of lasers for fiber optics. Just a few years ago, optical output powers of 1 to 2 mW were typical. Today, lasers are readily available with output powers of 10 mW to 40 mW or more. Much of the advance has been driven by the analog CATV market, always craving more optical output power, drives this advance. The current limit for output power is in the 100 mW range (+20 dBm) at 1550 nm. These very high-power lasers are typically used in continuous wave (CW) mode. In these cases, the modulation is imposed by an external modulator. The drawback of this setup is that the linewidth of the lasers can be as low as 100 kHz which corresponds to 0.000001 nm spectral width, as discussed earlier, very narrow spectral widths can trigger fiber nonlinearities such as SBS discussed in Chapter 3. The highest output lasers exhibit RIN's as low as -170 dB/Hz.

DWDM LASERS

With the rapid growth of DWDM technology in fiber optics, cooled DFB lasers are now available in a wide variety of precisely selected wavelengths. This precision can be achieved in two ways. First, the laser manufacturer has a number of different "recipes" for growing lasers that result in five to ten wavelength bands. The laser manufacture sorts the lasers and uses a built-in Peltier cooler to trim the wavelength to the exact required value. The user must then operate the temperature of the laser at a precise thermistor resistance value to ensure the correct wavelength. Table 5.2, shows the 45 ITU wavelengths or channels currently defined by the accepted standards. The wavelengths are generated by varying the optical frequency in 100 GHz or 0.1 THz increments. The frequencies (in THz) can then be converted to wavelength (in nm) by dividing them into the speed of light, 299,792.5 km/s. While this channel plan is usually referred to as 0.8 nm spacing, the channel spacing actually varies from 0.780 nm to 0.815 nm. Many manufacturers also refer to a 200 GHz channel spacing plan. In this case, every other value is used, resulting in a roughly 1.6 nm spacing.

Frequency (THz)	Wavelength (nm)	Frequency (THz)	Wavelength (nm)	Frequency (THz)	Wavelength (nm)
196.1	1528.77	194.6	1540.56	193.1	1552.52
196.0	1529.55	194.5	1541.35	193.0	1553.33
195.9	1530.33	194.4	1542.14	192.9	1554.13
195.8	1531.12	194.3	1542.94	192.8	1554.94
195.7	1531.90	194.2	1543.73	192.7	1555.75
195.6	1532.68	194.1	1544.53	192.6	1556.56
195.5	1533.47	194.0	1545.32	192.5	1557.36
195.4	1534.25	193.9	1546.12	192.4	1558.17
195.3	1535.04	193.8	1546.92	192.3	1558.98
195.2	1535.82	193.7	1547.72	192.2	1559.79
195.1	1536.61	193.6	1548.51	192.1	1560.61
195.0	1537.40	193.5	1549.32	192.0	1561.42
194.9	1538.19	193.4	1550.12	191.9	1562.23
194.8	1538.98	193.3	1550.92	191.8	1563.05
194.7	1539.77	193.2	1551.72	191.7	1563.86

Table 5.2: Standard ITU Wavelengths for DWDM, 100 GHz Spacing

Some users are extending the upper and lower bounds of the table in an effort to squeeze in more wavelengths per fiber. Others have cut the recommended minimum channel spacing in half, to 50 GHz, doubling the number of channels. This increases channel-to-channel crosstalk, increases the impact of some fiber nonlinearities such as FWM, and limits the maximum data rate per wavelength to OC-192, 10 Gb/s. OC-768, 40 Gb/s, simply will not easily fit in optical channels spaced at 50 GHz intervals.

MICROELECTROMECHANICAL SYSTEMS

Microelectromechanical systems (MEMS) combine electrical, mechanical, and optical components onto a single micro device or micro system. Ranging in size from micrometers to millimeters, MEMS use integrated circuit (IC) compatible batch-processing techniques during the fabrication process. MEMS process, sense, control, and actuate on the micro scale. They can function individually or in arrays to generate effects on the macro scale.

Texas Instruments (TI) designed one dramatic example of MEMS technology that will impact the general public. Digital light processing (DLP) technology leads other technologies for digital projectors in movie theaters. The DLP, fabricated using conventional silicon IC fabrication techniques, contains millions of precisely XY movable mirrors that create the image usually generated by film. Each mirror measures only 16 μm square and can be steered to an accuracy of less than 0.01°. The TI DLP is one of the earliest commercialized MEMS devices. Today, MEMS devices go into ink jet printers, read/write heads for magnetic storage devices, biosensors, and fiber optics.

A variety of newly developed etching techniques allow the creation of all sorts of mechanical devices on a silicon chip right along side conventional circuitry. The possibilities include levers, common in fiber optic applications, gears, and gimbals. The lever, most commonly used in fiber optics, moves by applying an electric field. Figure 5.22 shows an example of a MEMS device including a long lever arm on top of a VCSEL. The end of the arm consists of many dozens of layers that form a Bragg reflector. The air gap below the end of the lever arm varies by applying a voltage to the arm, which causes it to be attracted to the substrate.

Figure 5.22: MEMS
Device

(Photo courtesy of Bandwidth9.)

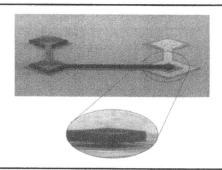

TUNABLE LASERS

Another new DWDM laser for use in systems carrying 100 wavelengths or more, the tunable laser, offers a convenient means of providing a single tunable backup transmitter that could replace any one of the 100 active transmitters in the DWDM system if they fail. Tunable lasers are also a key element in many advanced optical network architectures because they allow a data stream to be rapidly translated to a different wavelength. A variety of commercialized tunable laser architectures exist. Many tunable lasers in development base designs on VCSEL's and usually incorporate a moving element.

Figure 5.23: Tunable
VCSEL's

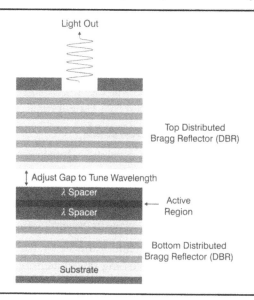

Figure 5.23 shows the structure of one type of tunable VCSEL. A variable gap tunes the wavelength. This gap might be varied using a MEMS device described earlier. The ideal tunable laser would be able to change wavelengths very quickly over a wide wavelength range. It would also have high output power that is constant as the wavelength changes. Figure 5.24 illustrates the response of voltage and wavelength in a tunable laser. Today's devices usually fall short on one or more of these factors. Nonetheless, tunable lasers will play an important part in advanced optical netwoks.

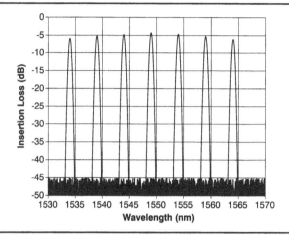

Figure 5.24: Tunable Laser

DFB LASERS

DFB lasers currently provide the most popular choice for DWDM LD's due their ability to stabilize the optical emission wavelength, their high output capability, narrow line width, and very low RIN. The goal for the optical emission wavelength stability of the DFB lasers is ±5 GHz for advanced DWDM systems. At a wavelength of 1550 nm, ±5 GHz corresponds to ±0.04 nm. The main factor that affects the wavelength of a DFB laser, other than the way it is manufactured, is the temperature of the laser. Most high-performance DFB lasers incorporate a TEC (thermoelectric cooler) and a thermistor into the laser package. The thermistor provides a highly accurate measure of the temperature of the laser chip and the TEC allows the laser chip to be heated or cooled by sending a DC current through the TEC. This allows the laser temperature to be precisely controlled over the life of the device thus maintaining wavelength stability. The ability to "lock" the wavelength of a DFB laser to a stable reference, such as the emission line of a gas, can lead to optical stability hundreds of times better than is required for even the most demanding DWDM system.

LASER SAFETY

While laser-based fiber optic products are in widespread use and present no hazard to personnel when handled properly, there are a few basic rules that must be observed to limit exposure to laser radiation. Laser radiation will damage eyesight under certain conditions. The following guidelines are important in laser safety:

- **Always** read the product data sheet and the laser safety label **before** applying power. Note the operating wavelength, optical output power, and safety classification.
- If safety goggles or other eye protection are used, be certain that the protection is effective at the wavelength(s) emitted by the device under test **before** applying power.
- **Always** connect a fiber to the output of the device **before** power is applied. Under no circumstances should the device ever be powered if the fiber output is unattached. If the device has a connector output, a connector that is connected to a fiber should be attached. This ensures that all light is confined within the fiber waveguide, virtually eliminating all potential hazard.
- **NEVER** look in the end of a fiber to see if light is coming out. Most fiber optic laser wavelengths (1310 nm and 1550 nm) are totally invisible to the unaided eye and will cause permanent damage. Shorter wavelength lasers (e.g. 780 nm) are visible and are potentially very damaging. Always use instrumentation, such as an optical power meter, to verify light output.

- **NEVER** look into the end of a fiber on a powered device with any sort of magnifying device. This includes microscopes, eye loupes, and magnifying glasses. This **will** cause a permanent, irreversible burn on the eye's retina. Always double check that power is disconnected before using such devices. If possible, completely disconnect the unit from any power source.

Laser safety classes are as follows in Table 5.3 Figure 5.25 illustrates typical laser warning labels used on the product packaging to identify laser-based products.:

Table 5.3: Laser Safety Classes

Class	Wavelength Range	Optical Power Accession Limits
I	180 nm to 10^6 nm	Varies with λ and exposure time.
IIa	400 nm to 710 nm	3.9×10^{-6} W (3.9 microwatts)
II	400 nm to 710 nm	1.0×10^{-3} (1.0 milliwatt)
IIIa	400 nm to 710 nm	5.0×10^{-3} (5.0 milliwatts)
IIIb	180 nm to 400 nm	Varies with λ and exposure time.
	400 nm to 10^6 nm	0.5 Watts

Figure 5.25: Typical Laser Warning Labels

COMPARISON OF LED'S AND LASER DIODES

Both types of light sources use the same key materials. GaAlAs (gallium aluminum arsenide) is commonly used for short-wavelength devices. Long-wavelength devices generally incorporate InGaAsP (indium gallium arsenide phosphide).

Linearity is an important characteristic to both types of light sources for some applications. Linearity represents the degree to which the optical output is directly proportional to the electrical current input. Most light sources give little or no attention to linearity, making them usable only for digital applications. Analog applications require close attention to linearity. Nonlinearity in LED's and lasers causes harmonic distortion in the analog signal that is transmitted over an analog fiber optic link.

LED's are generally more reliable than lasers, but both sources will degrade over time. This degradation can be caused by heat generated by the source and uneven current densities. In addition, LED's are easier to use than lasers. Lasers are temperature sensitive; the lasing threshold changes with the temperature. Thus, lasers require a method of stabilizing the threshold to achieve maximum performance. Often, a photodiode is used to monitor the light output on the rear facet of the laser. The current from the photodiode changes with variations in light output and provides feedback to adjust the laser drive current. Table 5.4 offers a quick comparison of some laser and LED characteristics.

Another issue in selecting LED's or lasers is the packaging. The main concern with packaging is how the light source couples with the fiber. One method uses a microlensed device. A small drop of epoxy is placed directly on the chip. The epoxy bead focuses the light in a uniform spot. The fiber, which usually has a smaller diameter than the bead, can be placed anywhere on the epoxy drop and will receive the same amount of optical energy. This allows for more efficient coupling of the fiber to the source. Pigtails use a

Parameter	Light-emitting Diode (LED)	Laser Diode (LD)
Output Power	Linearly proportional to drive current.	Proportional to current above the threshold.
Current	Drive Current: 50 to 100 mA Peak	Threshold Current: 5 to 40 mA
Coupled Power	Moderate	High
Bandwidth	Moderate	High
Wavelengths Available	0.66 to 1.65 μm	0.78 to 1.65 μm
Emission Spectrum	40 nm to 190 nm FWHM	0.00001 nm to 10 nm FWHM
Cost	$5 to $300	$5 to $3,000

Table 5.4: Comparison of Light Emitters

short length of fiber as part of the optical device. Pigtailing can improve coupling efficiency by bringing the fiber end closer to the emitting area of the chip. The light is coupled into the fiber before it has a chance to spread out.

LIGHT EMITTERS AS DETECTORS

All LED's and lasers have the ability to act as detectors, but a few perform this task much better than others. The key parameter to look for is very efficient coupling between the light emitter and the fiber. This allows good performance in both modes. It is also important that the LED's have consistent spectral characteristics. In ping-pong LED's or full-duplex LED's, the LED is used intermittently as a light emitter then as a light detector. In this way, information can be sent in either direction over the fiber. Ping-pong LED operation is illustrated in Figure 5.26. Some sophisticated techniques exist that combine this simple idea with time compression and an elaborate synchronizing protocol that appears to allow information to be sent both ways at the same time. In fact, information only travels in one direction at a time. The interchange back and forth is so quick that it seems to be simultaneous.

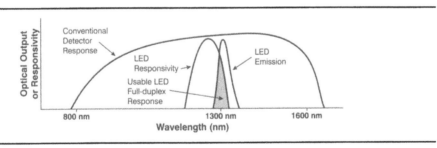

Figure 5.26: Ping-pong LED Operation

Now that we've discussed light emitters, it is time to move to the opto-electronic components at the other end of the fiber, light detectors.

Chapter Summary

- Light emitters convert electrical signals into corresponding light signals that can be injected into a fiber.
- Laser diodes (LD's) and light-emitting diodes (LED's) are the two types of light emitters in widespread use in modern fiber optic systems.
- LD's and LED's both use the principle of the p-n semiconductor junction found in transistors and diodes.
- Some characteristics that determine the usefulness of LD's and LED's are peak wavelength, spectral width, emission pattern, power, and speed.
- The two main types of LED's currently being used are surface-emitters and edge-emitters.
- Surface-emitters are the most economical of the two types of LED's, but they have low output power and are generally slower devices.

- Edge-emitters can produce high optical power levels and are generally faster than surface-emitters.
- A common trait of all LED types is that they turn on much faster than they turn off.
- The three main types of laser diode structures are the Fabry-Perot (FP), distributed feedback (DFB) and vertical cavity surface-emitting lasers (VCSEL).
- Chirp results from variations in the output wavelength of a laser diode.
- All lasers are susceptible to backreflections.
- The importance of controlling backreflections depends on the type of information being sent and the particular laser.
- Peltier coolers are specialized semiconductors that act like miniature heat pumps.
- Today, Peltier coolers are used in conjunction with very high-power lasers to remove heat from the laser chip and assure wavelength stability.
- Laser radiation can damage eyesight under certain conditions.
- LED's and LD's can survive and operate in extreme temperature ranges.
- LED's and LD's are usually driven with either a digital signal or an analog signal.
- Continuous wave lasers use external modulation to achieve output powers of 40 mW or more.
- DWDM lasers use precisely spaced wavelengths to achieve high channel counts in DWDM systems.
- Microelectromechanical systems combine electrical, mechanical, and optical components onto a single IC to process, sense, control, and actuate on a micro scale, or in arrays to generate effects on a macro scale.
- Tunable lasers provide the ability to adjust a laser to a specific DWDM wavelength, a feature that allows these devices to be used as back-up transmitters.
- Ping-pong, or full-duplex LED's are used intermittently as light emitters, then light detectors, enabling information to be sent both ways over the fiber.

Selected References and Additional Reading

Baack, Clemens. *Optical Wideband Transmission Systems.* Florida: CRC Press, Inc., 1986.

—. *Designer's Guide to Fiber Optics.* Harrisburg, PA: AMP, Incorporated, 1982.

Kopp, Greg and Laura A. Pagano. "Polarization Put in Perspective." *Photonics Spectra.* February 1995: 103-107.

Palladino, John R. *Fiber Optics: Communicating By Light.* Piscataway, NJ: Bellcore, 1990.

Rockwell, R. James, James F. Smith, and William J. Ertle. "Playing it Safe with Industrial Lasers." *Photonics Spectra.* February 1995: 118-124.

Sterling, Donald J. *Amp Technician's Guide to Fiber Optics*, 2nd Edition. New York: Delmar Publishers, 1993.

Yeh, Chai. *Handbook of Fiber Optics: Theory and Applications.* New York: Academic Press, Inc., 1990.

Zhang, Barry and Liangju Lu. "Isolators Protect Fiberoptic Systems and Optical Amplifiers." *Laser Focus World.* November 1998: 147-152.

LIGHT DETECTORS

Light detectors perform the opposite function of light emitters. Emitters, as we already know, are electro-optic devices. They convert electrical pulses into light pulses. Detectors are opto-electric devices. They convert the optical signals back into electrical impulses that are used by the receiving end of the fiber optic data, video, or audio link. The most common detector is the semiconductor photodiode, which produces current in response to incident light.

In an LED, the energy emitted during the recombination of electron-hole pairs is in the form of light. In a photodiode, the opposite phenomenon occurs. Light striking the photo-diode creates a current in the external circuit. Absorbed photons excite the electrons and the result is the creation of an electron-hole pair. For each electron-hole pair created, an electron is set flowing as current in the external circuit. Like light emitters, detectors operate based on the principle of the p-n junction. An incident photon striking the diode gives an electron in the valence band sufficient energy to move to the conduction band, creating a free electron and a hole. If the creation of these carriers occurs in a depleted region, the carriers will quickly separate and create a current. As they reach the edge of the depleted area, the electrical forces diminish and current ceases. While the p-n diodes are insufficient detectors for fiber optic systems, both PIN photodiodes and avalanche photo-diodes (APD's) are designed to compensate for the drawbacks of the p-n diode.

IMPORTANT DETECTOR PARAMETERS

Responsivity

The responsivity of a photodetector is the ratio of the current output to the light input. Other factors being equal, the higher the responsivity of the photodetector, the better the sensitivity of the receiver. Since responsivity varies with wavelength, it is specified either at the wavelength of peak responsivity or at a wavelength of interest. For most applications, responsivity is the most important characteristic of a detector because it defines the relationship between optical input and electrical output. The theoretical maximum responsivity is about 1.05 A/W at a wavelength of 1310 nm. Commercial InGaAs detectors provide typical responsivity of 0.8 to 0.9 A/W at a wavelength of 1310 nm.

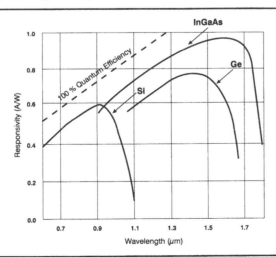

Figure 6.1: Typical Spectral Response of Various Detector Materials

The theoretical maximum responsivity of a photodetector occurs when the quantum efficiency of the detector is 100%. Responsivity and quantum efficiency (η) are related by:

Eq. 6.1

$$R = \frac{\eta \cdot \lambda}{1240}$$

Where:

R = Responsivity of the detector in Amps/Watt.
η = Quantum efficiency.
λ = Wavelength in nanometers.

Thus a detector at 1310 nm would have a theoretical maximum responsivity of 1.05 A/W and a detector at 850 nm would have a theoretical maximum responsivity of 0.68 A/W.

Quantum Efficiency

Quantum efficiency is the ratio of primary electron-hole pairs created by incident photons to the photons incident on the diode material. A quantum efficiency of 100% means that every absorbed photon creates an electron-hole pair. A typical quantum efficiency for a commercial detector is 70% to 90%. This quantum efficiency rating indicates that seven to nine out of ten photons create carriers. Factors that prevent the quantum efficiency from being 100% include coupling losses from the fiber to the detector, absorption of light in the p or n region, and leakage currents in the detector.

Capacitance

The capacitance of a detector is dependent upon the active area of the device and the reverse voltage across the device. A small active diameter allows for lower capacitance. However, as the active diameter decreases, it becomes harder to align the fiber to the detector. This is complicated by the fact that photodiode response is slower at the edges of the active area. If the edges of the detector are illuminated, a slow response component will be present, increasing edge jitter. It is important to illuminate only the center region of the active area to minimize this effect.

Photodiode capacitance decreases with increasing reverse voltage. Figure 6.2 shows a typical capacitance-voltage, or C-V curve, for a high-speed photodiode. This curve shows that as the reverse voltage is increased beyond 5 or 6 Volts, the decrease in capacitance becomes minimal. At this point the detector is said to be fully depleted. Higher reverse voltages also speed up the detector. Excessive reverse voltage may damage the detector or, in some cases, increase detector noise.

Figure 6.2:
Capacitance versus
Reverse Voltage

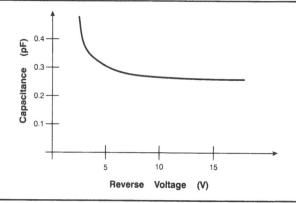

RESPONSE TIME

Response time represents the time needed for the photodiode to respond to optical inputs and produce an external current. The combination of the photodiode capacitance and the load resistance, along with the design of the photodiode, determines the response time. As with light emitters, detector response time is specified as rise time or fall time, and it is usually measured between 10% and 90% amplitude points. The response time of a diode relates to its usable bandwidth.

The design of the photodiode as well as its application parameters influence detector response time. For instance, the size of a detector (active area, usually expressed as a diameter in mm or μm) directly influences its capacitance. The applied reverse voltage decreases the capacitance and speeds response. The impedance that the detector operates into also affects the response time. The approximate -3 dB frequency of a detector is given in Equation 6.2.

$$\text{Eq. 6.2} \qquad f_{-3dB} = \frac{1}{(2 \cdot \pi \cdot R \cdot C)}$$

Where:

R = Impedance that the detector operates into.
C = Capacitance of the detector.

The 10-90% rise or fall time of the detector can also be estimated from Equation 6.3.

$$\text{Eq. 6.3} \qquad \tau = 2.2 \cdot R \cdot C$$

Equation 6.4 relates rise and fall time to -3 dB bandwidth.

$$\text{Eq. 6.4} \qquad \tau = \frac{0.35}{f_{-3dB}} \quad \text{or} \quad f_{-3dB} = \frac{0.35}{\tau}$$

Consider a high-speed detector with a capacitance (C) of 0.5 pF operating into an impedance (R) of 50 Ohms. This detector would exhibit a rise time of 55 ps and a -3 dB frequency of 6.4 GHz. Keep in mind that other detector factors could limit the performance to lower values. The calculated values should be considered the best that could be achieved.

Dark Current

This is one of the worst terms ever conceived for a phenomenon. It implies that the detector somehow manages to put out a current when there is no light. Actually, a small current, called dark current, flows through the detector in the absence of light because of the intrinsic resistance of the detector and the applied reverse voltage. The voltage acting on the bulk resistance of the detector causes this dark current, which is very temperature sensitive and may double every 5°C to 10°C. Dark current contributes to the detector noise and also creates difficulties for DC coupled amplifier stages.

Edge Effect

Detectors only provide fast response in their center region which is an often overlooked property of the detector. The outer regions of the detector exhibit a phenomenon known aptly enough as the edge effect. This phenomenon affects the detector's response in two ways. First, the detector edge has a higher responsivity compared to the center. This causes problems when aligning a fiber to the detector. If a continuous wave light source is used, the operator will almost always be fooled into aligning the fiber to the edge of the detector, not the center, because of the higher responsivity.

So why is higher responsivity a bad thing? Because the second implication of the edge effect is that the response is often significantly slower at the edge compared to the central region. For that reason, it is always a good idea to align detectors to fiber using a high frequency square wave light source and an oscilloscope to look for the cleanest edges. Usually a frequency of 1 MHz or higher is adequate to ensure that the detector edge effect is avoided. Figure 6.3a shows the response of a detector that is properly aligned to the center of the detector. Figure 6.3b illustrates the response of a detector that is improperly aligned and shows the ill effects of the slow edge region of the detector.

Figure 6.3: Edge Effect

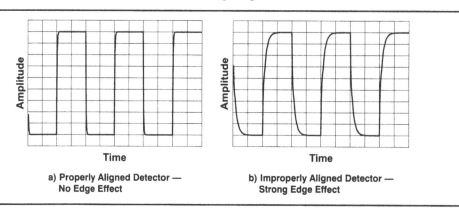

a) Properly Aligned Detector — No Edge Effect

b) Improperly Aligned Detector — Strong Edge Effect

Linearity & Backreflection

All PIN diodes are inherently linear devices. However, for the most demanding applications, such as multichannel CATV links, special care must be taken to reduce distortion to very low levels. These so-called analog PIN detectors often have distortion products below -60 dB. Another factor that is very important for analog applications is the backreflection of the detector. Generally, the fiber is coupled to the detector at a perpendicular angle. For low-backreflection detectors, the detector may be tilted by 7° to 10° as shown in Figure 6.4.

Figure 6.4: Low Backreflection Detector Alignment

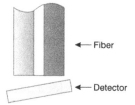

Fiber

Detector

Aligning a fiber to a low-backreflection, low-distortion PIN diode can be trickier than aligning a fiber to a laser diode. The manufacturer has to align the fiber for cleanest response (to eliminate the edge effect), lowest backreflection, highest responsivity, and lowest distortion. Typical test fixtures used for this detector alignment display all four parameters at once.

Noise

Noise is an ever-present phenomenon that limits a detector's performance. It is any electrical or optical energy other than the signal itself. Noise appears in all elements of a communication system; however, it is usually most critical to the receiver because the receiver is trying to interpret an already weak signal. The same noise in a transmitter is usually insignificant since the signal at the transmitter is much stronger than the attenuated signal that the receiver picks up. Shot noise occurs because the process of creating the current is a set of discrete occurrences rather than a continuous flow. As more or less electron-hole pairs are created, the current fluctuates, creating shot noise. Shot noise occurs when no light falls on the detector. Even without light, a small current, dark current, is thermally generated. Noise increases with current and bandwidth. When dark current only is present, noise is at a minimum, but it increases with the current resulting from optical input.

A second type of noise, thermal noise, arises from fluctuations in the load resistance of the detector. The electrons in the resistor are not stationary, and their thermal energy allows them to move about. At any given moment, the net movement toward one electrode or the other generates random currents that add to and distort the signal current from the photodiode.

Shot noise and thermal noise exist in the receiver independent of the arriving optical power. They are a result of matter itself, but they can be minimized. A rule of thumb is that the signal power should be ten times that of the noise or the signal will not be adequately detected. This signal quality can be expressed as a signal-to-noise ratio (SNR). A large SNR means that the signal is much larger than the noise. Different applications require different signal-to-noise ratios. This signal-to-noise ratio is treated further in Chapter 10, "System Design Considerations."

PIN PHOTODIODE

A p-n diode's deficiencies are related to the fact that the depletion area (active detection area) is small; many electron-hole pairs recombine before they can create a current in the external circuit. In the PIN photodiode, the depleted region is made as large as possible. A lightly doped intrinsic layer separates the more heavily doped p-types and n-types. The diode's name comes from the layering of these materials Positive, Intrinsic, Negative - PIN.

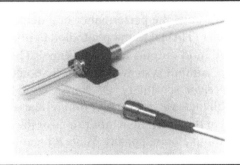

Figure 6.5: Low Backreflection Photodiodes

(Photo courtesy of PD-LD, Inc.)

In the absence of light, PIN photodiodes behave electrically just like an ordinary rectifier diode. If forward biased, they conduct large amounts of current. The forward turn-on voltage is related to the energy gap of the detector. For silicon, this energy gap is around 1.1 eV (electron Volts). For InGaAs, the energy gap is 0.77 eV and for germanium, the energy gap is around 0.65 eV. Figure 6.6 shows the cross section and operation of a PIN photodiode.

PIN detectors operate in two modes: photovoltaic and photoconductive. In the photovoltaic mode, no bias is applied to the detector. In that case the detector will be very slow, and the detector output is a voltage that is approximately logarithmic to the input light level. Real-world fiber optic receivers never use the photovoltaic mode. In the photoconductive mode, the detector is reverse biased. The output in this case is a current that is very linear with the input light power. PIN detectors can be linear over seven or more decades of input light intensity.

Figure 6.6: Cross-section and Operation of a PIN Photodiode

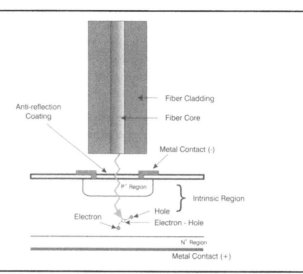

IDP DETECTORS

An alternative to the PIN photodiode is known as an integrated detector/preamplifier (IDP). Noise that limits the operation of the receiver can occur between the diode and the first receiver stage. To reduce this noise an IDP may be used. An IDP is an integrated circuit that has both a detector and a transimpedance amplifier (see discussion below). Characteristic specifications for IDP's are similar to those of PIN detectors or APD's. One difference is that the output of an IDP is voltage, so responsivity is specified in Volts/Watt (V/W). Typical responsivity for an IDP is 4000 V/W.

Transimpedance Amplifier

These devices are used to improve the performance of a detector in a fiber optic system. The transimpedance amplifier (TIA) takes the photodiode output current, multiplies it by the transimpedance gain, A_z, and outputs a voltage signal. The TIA must have sufficient bandwidth to handle the maximum system data rate. It should also have high gain and low noise to maximize receiver sensitivity. Of course, as the bandwidth or the gain increases, the noise floor will increase. A TIA with differential outputs is very beneficial to receiver sensitivity because of the differential input stage of the decision circuit. This effectively adds 3 dB of gain without a noise penalty. Table 6.1 shows some typical characteristics of a transimpedance amplifier. TIA's may be used with PIN or APD detectors.

Table 6.1: TIA Characteristics

Transimpedance Gain	10,000 Ω
Bandwidth	150 MHz
Input Noise Current Spectral Density	2.9 pA/$\sqrt{Hz}$
Output Stage	Single-ended

AVALANCHE PHOTODIODE

Another type of detector is the avalanche photodiode (APD). For PIN diodes, each absorbed photon ideally creates one electron-hole pair that sets one electron flowing in the external circuit. In this one-to-one relation, PIN photodiodes resemble LED's. Lasers offer a higher-than-one-to-one ratio of photons to carriers, and current: one primary carrier can result in the emission of several photons. By this comparison, APD's are like lasers. The primary carriers, the free electrons and holes created by absorbed photons, accelerate, gaining several electron Volts of kinetic energy. A collision of these fast carriers with neutral atoms causes the accelerated carriers to use some of their own energy to help the bound electrons break out of the valence shell. Free electron-hole pairs, called secondary carriers, appear. Collision ionization is the name for the process that creates these secondary carriers. As primary carriers create secondary carriers, the secondary carriers themselves accelerate and create new carriers. Collectively, this process is known as photomultiplication. Typical multiplication ranges in the tens and hundreds. A multiplication factor of eighty means that, on average, eighty external electrons flow for every photon of light absorbed. This factor is a statistical average — the actual number of electrons generated per photon absorbed may be more or less than the multiplication factor.

APD's are of interest because of the inherent gain that they provide. Because they include gain, the electronics amplifier chain can be simplified. However, as in most engineering trade-offs, there is a price. For APD's the price is heavy. APD's require high-voltage power supplies for their operation. The voltage can range from 30 or 70 Volts for InGaAs APD's to over 300 Volts for Si APD's. This adds circuit complexity. Also, APD's are very temperature sensitive, further increasing circuit complexity. In general, APD's are only useful for digital systems because they possess very poor linearity. Because of the added circuit complexity and the high voltages that the parts are subjected to, APD's are always less reliable than PIN detectors. This, added to the fact that at lower data rates, PIN detector-based receivers can almost match the performance of APD-based receivers, makes PIN detectors the first choice for most deployed low-speed systems. At multigigabit data rates, however, APD's rule supreme.

APD-TIA DETECTORS

Like IDP's discussed earlier, APD's are often integrated with amplifiers as a single component. This is especially important for applications at high speeds, above 1 Gb/s. The capacitance of the detector and the first amplifier input are usually the dominant factors that determine how fast and sensitive the complete receiver will be. By integrating the APD and the TIA into one package, the capacitance associated with the connection from the APD output to the amplifier input is minimized. Because APD's are temperature sensitive, many manufacturers also include a thermistor in the package to monitor the temperature of the APD. Currently, APD-TIA combinations are available to handle data rates to 10 Gb/s and surely 40 Gb/s cannot be far away.

OTHER DETECTOR ENHANCEMENT TECHNIQUES

We will discuss erbium-doped fiber amplifiers (EDFA's) and semiconductor optical amplifiers (SOA's) in detail in Chapter 7. It is appropriate to mention here that EDFA's and SOA's are increasingly used in front of a receiver to enhance its sensitivity. This is especially true at data rates of 10 Gb/s and above. In this case, the EDFA or SOA first provides optical gain, the APD converts the optical signal to electrons, adding more gain, followed by a TIA that provides additional electrical gain.

COMPARISON OF PIN PHOTODIODES AND APD'S

Table 6.2 offers a comparison of PIN photodiodes and APD's.

Parameter	PIN Photodiodes	APD's
Construction Materials	Silicon, Germanium, InGaAs	Silicon, Germanium, InGaAs
Bandwidth	DC to 40+ GHz	DC to 40+ GHz
Wavelength	0.6 to 1.8 μm	0.6 to 1.8 μm
Conversion Efficiency	0.5 to 1.0 Amps/Watt	0.5 to 100 Amps/Watt
Support Circuitry Required	None	High Voltage and Temperature Stabilization
Cost (Fiber Ready)	$1 to $500	$100 to $2000

Table 6.2: Comparison of PIN Photodiodes and APD's

As with light emitters, coupling the detector to the fiber is an important consideration. Detectors are packaged in the same ways as light emitters, using microlensed devices or pigtailing the detector.

Chapter Summary

- Light detectors enable an optical signal to be converted back into electrical impulses that are used by the receiving end of the fiber optic data, video, or audio link.
- The most common detector is the semiconductor photodiode which produces current in response to incident light.
- The responsivity of a photodetector is the ratio of the current output to the light input.
- Quantum efficiency is the ratio of primary electron-hole pairs created by incident photons to the number of photons incident on the diode material.
- The theoretical maximum responsivity of a photodetector occurs when the quantum efficiency of the detector is 100%.
- The capacitance of a detector is dependent upon the active area of the device, the design of the junction, and the reverse voltage across the device.
- Photodiode capacitance decreases with increasing reverse voltage.
- Dark current refers to the flow of current through a detector in the absence of light.
- The combination of the photodiode capacitance and the load resistance, along with the design of the photodiode, sets the time needed for the photodiode to respond to optical inputs and produce an external current.
- Noise (shot and thermal) is an ever-present phenomenon that limits a detector's and a system's performance.
- PIN photodiodes and avalanche photodiodes (APD's) are designed to compensate for the drawbacks of the p-n diode.
- PIN photodiodes behave electrically, like an ordinary rectifier diode, in the absence of light.
- An IDP is an integrated circuit that has both a detector and a transimpedance amplifier.
- In APD's, collision ionization is the process that creates secondary carriers.
- APD's require high-voltage power supplies for their operation, ranging from 30 to 70 Volts for InGaAs APD's, to over 300 Volts for Si APD's.
- Because of the many shortcomings of APD's, PIN detectors are the first choice for most systems. Selecting a detector depends on the application involved, cost, and bandwidth.
- An APD-TIA is an avalanche photodiode and a transimpedance amplifier integrated into one package to minimize input capacitance.

Selected References and Additional Reading

Baack, Clemens. *Optical Wideband Transmission Systems.* Florida: CRC Press, Inc., 1986.

Designer's Guide to Fiber Optics. Harrisburg, PA: AMP, Incorporated, 1982.

Hentschel, Christian. *Fiber Optics Handbook.* 2nd edition. Germany: Hewlett-Packard Company, 1988.

Palladino, John R. *Fiber Optics: Communicating By Light.* Piscataway, NJ: Bellcore, 1990.

Yeh, Chai. *Handbook of Fiber Optics: Theory and Applications.* New York: Academic Press, Inc., 1990.

OTHER ACTIVE DEVICES

SEMICONDUCTOR OPTICAL AMPLIFIERS

Semiconductor optical amplifiers (SOA's) are essentially laser diodes, without end mirrors, which have fiber attached to both ends. They amplify any optical signal that comes from either fiber and transmit an amplified version of the signal out the second fiber. An SOA can be constructed in a small package, and they work for 1310 nm or 1550 nm systems. In addition, SOA's transmit bidirectionally, making the reduced size of the device an advantage over regenerators or EDFA's. Drawbacks to SOA's include high-coupling losses, polarization dependence, and a higher noise figure. Figure 7.1 illustrates the basics of a semiconductor optical amplifier.

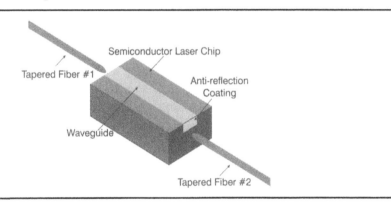

Figure 7.1:
Semiconductor
Optical Amplifier

Modern optical networks utilize SOA's in a number of ways including:

Power Booster: Many tunable laser designs output low optical power levels and must be immediately followed by an optical amplifier, usually either an SOA or EDFA.

In-Line Amplifier: Allows signals to be amplified within the signal path.

Wavelength Conversion: Covered later in this chapter, this involves changing the wavelength of an optical signal.

Receiver Preamplifier: SOA's can be placed in front of detectors to enhance sensitivity.

ERBIUM-DOPED FIBER AMPLIFIERS

The breakthrough development of erbium-doped fiber amplifiers (EDFA's) allows long-haul DWDM systems to flourish. Before EDFA's, DWDM systems required electronic repeaters at intermediate points about 100 km apart. Repeater's are essentially a fiber optic receiver connected directly to a fiber optic transmitter. EDFA's allow the transmission of optical signals over longer distances without the need for repeaters. EDFA's also allow the DWDM system to be upgraded by adding additional sources to different wavelengths and combining them onto a single fiber using a DWDM multiplexer.

At the heart of the EDFA, the fiber is doped with erbium, a rare earth element that has the appropriate energy levels in its atomic structure for amplifying light at 1550 nm. A 980 nm or 1480 nm "pump" laser injects energy into the erbium-doped fiber. When a weak signal at 1550 nm enters the fiber, the light stimulates the erbium atoms to release their stored energy as additional 1550 nm light. This process continues as the signal passes down the fiber, growing stronger and stronger as it goes.

Figure 7.2: Two-stage EDFA with Mid-stage Access

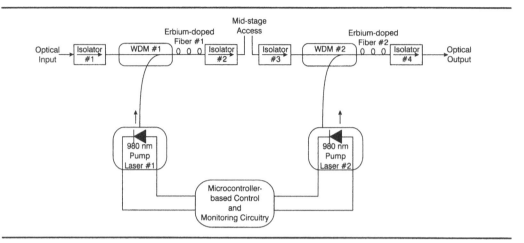

Figure 7.2 shows a two-stage EDFA with mid-stage access. In this case, two simple single-stage EDFA's are packaged together. The output of the first stage EDFA and the input of the second stage EDFA are brought out to the user. Mid-stage access is important in very high performance fiber optic systems. These systems often require the periodic use of elements such as dispersion-compensating fiber (DCF) in order to reduce the overall dispersion. A high insertion loss of 10 dB or more makes DCF problematic. Placing the DCF at the mid-stage access point of the two-stage EDFA reduces detrimental effects on the system. The user still realizes significant gain through the EDFA, even though a high optical loss piece of DCF has been added.

A typical EDFA operates as follows (see Figure 7.2): The optical input first passes through optical isolator, Isolator #1. An optical isolator is a device that only allows light to pass from left to right. Next the light passes through WDM #1. WDM #1 provides a means of injecting the 980 nm pump wavelength into the first length of erbium-doped fiber. WDM #1 also allows the optical input signal to be coupled into the erbium-doped fiber with minimal optical loss. The erbium-doped optical fiber is usually tens of meters long. The 980 nm energy pumps the erbium atoms into an excited state that decay slowly. When light in the 1550 nm band travels through the erbium-doped fiber it causes stimulated emission of radiation, much like in a laser. In this way, the 1550 nm optical input signal gains strength. The output of the erbium-doped fiber then goes through optical isolator, Isolator #2. The output of Isolator #2 is made available to the user. Typically some sort of dispersion compensating device will be connected at the mid-stage access point. The light then travels through Isolator #3 and the WDM #2. WDM #2 couples additional 980 nm energy from a second pump laser into the other end of a second length of erbium-doped fiber, increasing gain and output power. Finally the light travels through Isolator #4.

Photons amplify the signal, a benefit of EDFA's that avoids almost all active components. The output power from an EDFA can be quite large, requiring fewer amplifiers in a given system design. Furthermore, data rate independence in EDFA's means a system upgrade requires changing only the launch/receive terminals. EDFA's already in place will work fine for future data rate upgrades. The most basic EDFA design amplifies light over a fairly narrow, 12 nm, band. The addition of gain equalization filters, can increase the band to more than 25 nm. Other exotic doped fibers increase the amplification band to 40 nm.

The performance characteristics of EDFA's make them useful in long-haul, high data rate fiber optic communications systems and CATV delivery systems. Long-haul systems need amplifiers because of the lengths of fiber used. CATV applications often need to split a signal to several fibers, and EDFA's boost the signal before and after the fiber splits. In general, there are four major applications for optical fiber amplifiers: power amplifier/ booster, in-line amplifier, preamplifier or loss compensation for optical networks. These applications are described in more detail below.

Figure 7.3: Typical Packaged EDFA

(Photo courtesy of Nortel Networks, Optical Components.)

Power Amplifier/Booster

Figure 7.4 illustrates the first three applications for optical amplifiers. Power amplifiers (also called booster amplifiers) are placed directly after the optical transmitter. In this application, EDFA's need to be able to take a large signal input and provide maximum possible output level. Small signal response is not as important because the direct transmitter output is usually -10 dBm or higher. The noise added by the amplifier at this point is also not as critical because the incoming signal has a large signal-to-noise ratio (SNR).

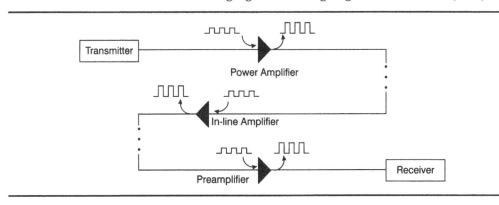

Figure 7.4: Three Applications for an Erbium-doped Fiber Amplifier

In-line Amplifier

The in-line amplifier or in-line repeater, takes a small input signal and boosts it for retransmission down the fiber. It is important to control the small signal performance and the noise added by the EDFA. Noise added by amplifiers in series will limit the system length.

Preamplifier

In the past, a receiver sensitivity of -30 dBm at 622 Mb/s might have been adequate, but today's customers demand sensitivities of -40 dBm or -45 dBm. This performance can be achieved by placing an optical amplifier prior to the receiver. Boosting the signal at this point presents a much larger signal into the receiver, thus easing the demands of the receiver design. For this application, the noise added by the EDFA is critical. The amplifier must add a minimum of noise to maximize the received SNR.

Loss Compensation in Optical Networks

This application, illustrated in Figure 7.5, inserts the EDFA before an 8 x 1 optical splitter. The optical splitter has a nominal optical insertion loss of 10 dB. If the transmitter has an optical output of +10 dBm, the splitter outputs without the EDFA would be 0 dBm. In digital applications, this output power suffices; however, in analog CATV applications, this output represents the minimum acceptable received power. Inserting an EDFA before the optical splitter, increases the power to perhaps +19 dBm, allowing each of the eight output legs to provide +9 dBm, nearly equal the original transmitter power.

Figure 7.5: Loss Compensation in Optical Networks

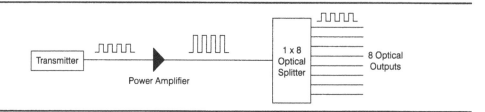

Wideband EDFA's

Optical communication systems which may carry 100 or more optical wavelengths require an increase in the bandwidth of the optical amplifier to nearly 80 nm. Usually this involves a hybrid optical amplifier, consisting of two separate optical amplifiers, one for the lower 40 nm band and the second for the upper 40 nm band. Figure 7.6 shows the optical gain spectrum of a hybrid optical amplifier. The solid lines show the response of two individual amplifier sections. The dotted line, which has been increased by 1 dB for clarity, shows the response of the combined hybrid amplifier.

Figure 7.6: Optical Gain Spectrum of a Hybrid Optical Amplifier

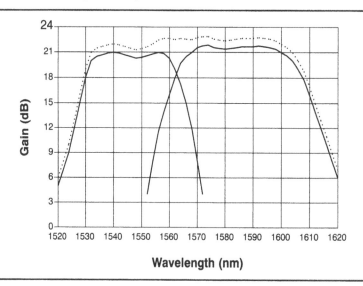

Figure 7.7 shows various optical amplifier topologies and the bandwidth performance level that they can achieve.

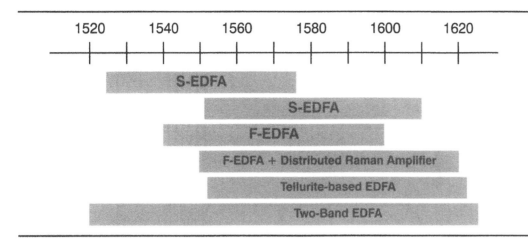

Figure 7.7: Wideband EDFA/Raman Amplifiers Achieved by 1998

A number of EDFA types have been constructed to work as wideband amplifiers in the 1.5 micron telecommunications window. These devices employ several types of fiber components and gain equalizers, as well as varying pumping schemes. Using partially gain-flattened wideband EDFA and Raman amplification in the transmission fiber, an extremely large bandwidth of 67 nm (1549-1616 nm) can be obtained. Figure 7.8 illustrates this.

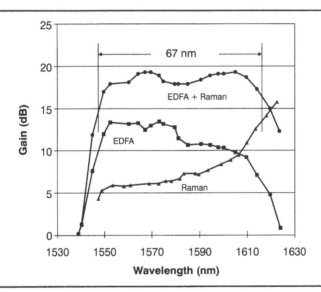

Figure 7.8: Gain of EDFA plus Raman Optical Amplifier Showing 67 nm Bandwidth

RAMAN AMPLIFIERS

The Raman optical amplifier differs in principle from EDFA's or conventional lasers. In EDFA's and conventional lasers, atoms are pumped to a high energy state and then drop to a lower state, releasing their energy, when a suitable wavelength photon passes nearby. Raman optical amplifiers utilize stimulated Raman scattering (SRS) to create optical gain. Initially SRS was considered to detrimental to high channel count DWDM systems. Figure 7.9 shows the typical transmit spectrum of a six channel DWDM system in the 1550 nm window. Note that all six wavelengths have approximately the same amplitudes.

Figure 7.9: Sample DWDM Transmit Spectrum, Six Wavelengths

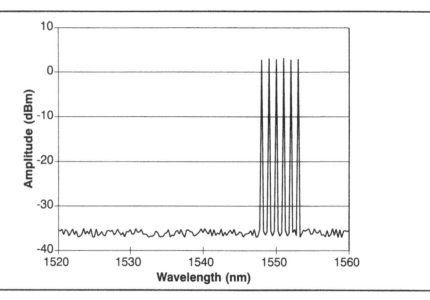

Notice the differences in Figure 7.10. First, the noise background has increased. The amplitudes of the six wavelengths are now very different. The lower wavelengths have a smaller amplitude than the upper wavelengths. The SRS effect has robbed energy from the lower wavelengths and fed that energy to the upper wavelengths.

Figure 7.10: Received Spectrum After SRS in a Long Fiber

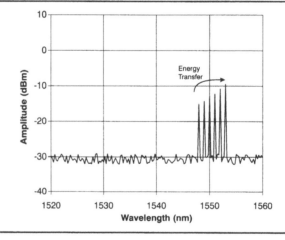

A Raman optical amplifier is not a "black box" like an EDFA. It consists of little more than a high-power pump laser, usually called a Raman laser, and a WDM or directional coupler. The optical amplification occurs in the transmission fiber itself, distributed along the transmission path. Optical signals are amplified up to 10 dB in the network optical fiber. The Raman optical amplifiers have wide gain bandwidth (up to 100 nm). They can use any installed transmission optical fiber (single-mode optical fiber, TrueWave, etc.). In effect, they reduce the effective span loss to improve noise performance by boosting the optical signal in transit. They can be combined with erbium-doped fiber amplifiers to very wide optical gain flattened bandwidth. Figure 7.11 shows the topology of a typical Raman optical amplifier. The pump laser and circulator comprise the two key elements of the Raman optical amplifier. The pump laser, in this case, has a wavelength of 1535 nm. The circulator provides a convenient means of injecting light backwards in to the transmission path with minimal optical loss.

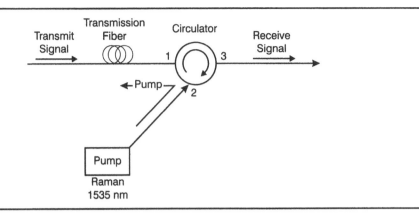

Figure 7.11: Typical Raman Amplifier Configuration

Figure 7.12 shows the optical spectrum of a forward-pumped Raman optical amplifier. In this case, the pump laser is injected at the transmit end rather than the receive end as shown in Figure 7.11. The pump laser has a wavelength of 1535 nm. As is usually the case, the amplitude of the pump laser is much greater than the data signals.

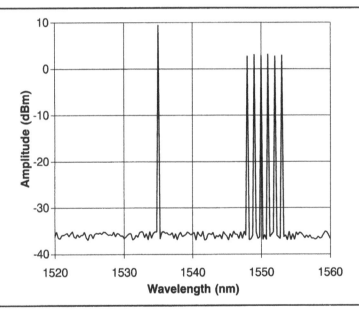

Figure 7.12: Raman Amplifier Example — Transmitted Spectrum

Figure 7.13 shows the received signal after the same length of fiber used in the SRS example above. While the amplitude of the pump laser is significantly decreased, the amplitude of the six data signals is now much stronger and they all have roughly equal amplitudes. In this case, a great deal of energy was robbed from the 1535 nm pump laser signal and redistributed to the six data signals.

Figure 7.13: Raman Amplifier Example — Received Spectrum

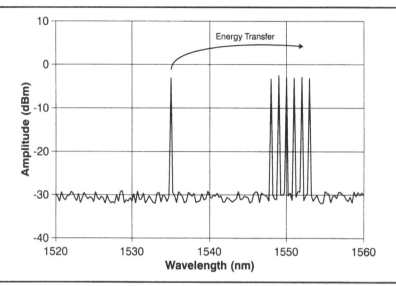

EXTERNAL MODULATORS

When data rates were in the low gigabit range and transmission distances were less than 100 km or so, most fiber optic transmitters used directly modulated lasers. However as data rates and span lengths grew, wavelength chirp, caused by turning a laser on and off, limited data rates. The wavelength chirp widened the effective spectral width of the laser which causes dispersion problems as we have seen in earlier chapters. A laser source with no wavelength chirp and a narrow linewidth provide one solution to the problem. This solution took the form of external modulation which allows the laser to be turned on continuously; the modulation is accomplished outside of the laser cavity.

An external modulator, which functions much like an electrically activated shutter, modulates of the light. As analog devices, external modulators allow the amount of light passed to vary from some maximum amount (P_{MAX}) to some minimum amount (P_{MIN}). Other key terms related to external modulators include:

Vπ: This is the voltage required to take the response function through ½ cycle or 180°.

Bias Point: The DC point around which the modulation signal swings.

Insertion Loss: The amount of loss from the light injected by the laser at the peak of the waveform. This usually amounts to 3-5 dB. Keep in mind that operating at the usual bias point will introduce an additional 3 dB of loss for a total insertion loss of 6 to 8 dB. (See Figure 7.18 for details.)

P$_{MIN}$: The minimum light output from the external modulator. Usually about 5% of the maximum value.

P$_{MAX}$: The maximum light output from the external modulator. Usually 3 to 5 dB less than the laser input.

P$_{AVG}$: The average light out of the external modulator. Usually 3 dB less than P_{MAX} if driven by a 50% duty cycle waveform.

Lithium Niobate Amplitude and Phase Modulators

The popularity of lithium niobate ($LiNbO_3$) as a material used in external modulators results from its low optical loss and high electro-optic coefficient. This coefficient refers to the electro-optic effect, which occurs in some materials such as lithium niobate, in which the refractive index of the material changes in response to an applied electric field. As we learned in earlier chapters, the refractive index of the material causes light to travel at a speed inversely proportional to the refractive index of the material. Thus if we could suddenly increase the refractive index of a material, we would slow the light beam down and vice versa.

Figure 7.14 shows the block diagram of a typical external modulator. The input light enters the external modulator via the input fiber. The light is first split into two fibers using an optical splitter. The top fiber path travels through a length of $LiNbO_3$ crystal. The light in the bottom fiber experiences a fixed delay. After the light travels through the lithium niobate crystal and the fixed length of fiber, an optical combiner merges the two fiber paths. The light travels through identical path legs.

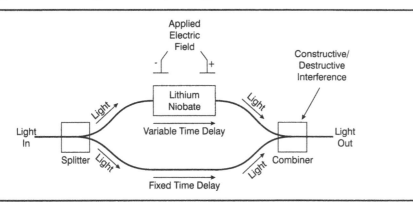

Figure 7.14: Basic Lithium Niobate ($LiNbO_3$) Optical Modulator

The speed at which light travels through any medium depends on the refractive index of the material. A higher refractive index results in slower light speed. $LiNbO_3$ exhibits the electro-optic effect. By applying an electric field to the material, its refractive index changes. We now see that if the time delay through the fixed fiber and the $LiNbO_3$ crystal is equal, the light will be in phase when it reaches the output optical combiner. Referring the nature of light as described in Chapter 2, we see that since the light in both legs are in phase, they will constructively add to form the maximum possible output. The refractive index and the speed of light change as the applied voltage changes. When the speed changes enough to delay the light by half of one wavelength, the light will be out of phase when it reaches the output 3 dB coupler. Now the light will destructively add yielding a minimum possible output.

In order to make a lithium niobate substrate suitable for use in fiber optic devices, one must first build a waveguide in the substrate. As with optical fiber itself, this is accomplished by introducing dopant materials into the area that will become the waveguide. Doping raises the refractive index of the waveguide relative to the surrounding substrate while maintaining optical transparency. Once accomplished, the waveguide will contain the light by the principles of total internal reflection. If the dimensions of the waveguide remain consistent with the dimensions of the core of a single-mode fiber, about nine microns in diameter, then light will efficiently couple into and out of the waveguide. This basic design proves useful in a fiber optic system.

Figure 7.15 shows the simplest type of external modulator, a phase modulator. The phase modulator has a single optical input of polarization maintaining (PM) fiber and a single optical output of PM or single-mode (SM) fiber. In a simple phase modulator, two electrodes surround the waveguide. The bottom electrode is grounded while the top electrode is driven by an outside voltage signal. As the voltage on the top electrode changes, the refractive index of the waveguide changes accordingly, alternating the light as the refractive index rises and falls. While this modulates the phase of the light, the output intensity remains unchanged. Recalling the discussion on fiber nonlinearities, this function overcomes SBS. The SBS threshold can increase by as much as 10 dB because phase modulating the light effectively spreads out the optical energy.

Figure 7.15: Simple Phase Modulator

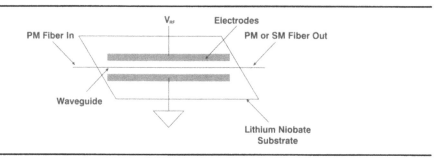

Figure 7.16 illustrates a slightly more internally complex device. It has the same input and output fiber setup as the simple phase modulator. However, after the light enters the lithium niobate waveguide, it optically splits into two paths using a fiber optic coupler designed into the substrate. These two paths travel for a distance and then recombine using another fiber optic coupler. If the light waves are in phase, they will add constructively to produce a large output on the output leg. If they are out of phase, destructive interference yields little or no output. The two paths of light travel through sets of electrodes arranged so that they have opposite effects on the two paths. By applying an external voltage, the refractive index of one path will rise while the refractive index of the other path falls. This causes the output optical amplitude to vary as the light from the two paths moves from constructive addition to destructive interference.

Figure 7.16: Single Output Intensity Modulator

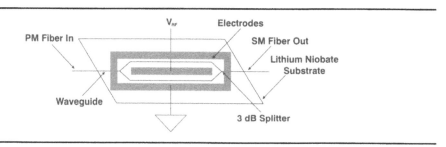

A third type of external modulator, illustrated in Figure 7.17, resembles the modulator shown in Figure 7.16. In this case, a 3 dB coupler forms at the output, giving two output fibers rather than one. The light amplitude of the two output legs will move opposite of each other. When the light level of one leg increases, the light level of the other leg decreases. The dual output modulator, which provides two out of phase outputs works best in analog drive situations.

Figure 7.17: Dual Output Intensity Modulator

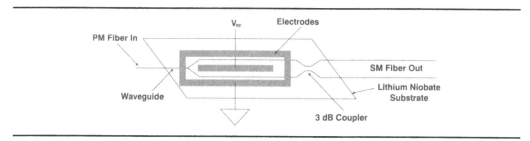

Figure 7.18 shows the typical raised sine function response of the dual output intensity modulator. In this case, the modulator operates around zero Volts bias. At zero Volts bias, the output intensity of both output legs is equal. As the applied voltage increases slightly, the intensity of output 2 increases, while the intensity of output 1 decreases. This continues until the voltage reaches $V_{\pi/2}$. At that point, the intensity of output 2 will be at a maximum and the intensity of output 1 will be at a minimum. This sine function response repeats as the applied voltage increases or decreases. Usually, modulator designers exploit the response nearest zero Volts bias.

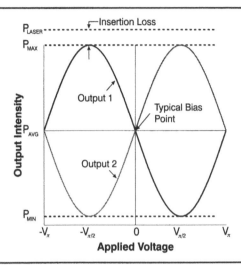

Figure 7.18: Dual Output External Modulator Response

Digital Operation

In the simple applications, an external modulator transmits a digital data stream. The drive voltage is toggled between $-V_{\pi/2}$ and $V_{\pi/2}$. This causes the output intensity to swing from maximum to minimum utilizing maximum modulation depth.

Analog Operation

External modulators may also be used to transmit analog signals. In this case, extensive stabilization and linearization may be required. Stabilizing the bias point at exactly the 50% point minimizes that second-order distortion. However, a third-order distortion remains. A small drive signal may yield a response that does not require linearization. For CATV applications sending 80 or 110 channels, considerable care must be taken to predistort the signal to remove the third-order distortion effects.

WAVELENGTH CONVERTERS

As the name implies, wavelength converters take an input wavelength and converts it to a different output wavelength. This is useful in many proposed "all-optical" network topologies where switching occurs at the optical level. In Figure 7.19, λ_s denotes the input signal wavelength, λ_c is the converted wavelength, and λ_p is the pump wavelength.

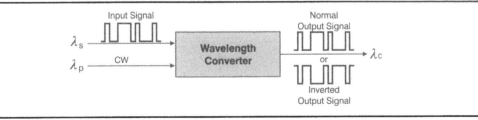

Figure 7.19: Wavelength Converter

An ideal wavelength converter would be transparent to bit rates and signal formats, have fast setup time of output wavelength, convert to both shorter and longer wavelengths, and possess a low-chirp output signal with high extinction ratio and large signal-to-noise ratio (SNR).

Wavelength Conversion Technologies

Wavelength conversion techniques can be broadly classified into two types: optoelectronic wavelength conversion, in which the optical signal must first be converted into an electronic signal, and all-optical wavelength conversion, where the signal remains in the optical domain. All-optical conversion techniques may be subdivided into techniques that employ coherent effects and techniques that use cross modulation. All-optical techniques are the most interesting since they are the only techniques that will be transparent to bit rates and signal formats. Some all-optical techniques provide a converted output signal that is the same as the input signal, while others yield an inverted output signal.

WAVELENGTH LOCKER

Narrow-spaced DWDM systems require lasers at very precise wavelengths. These systems often employ devices called wavelength lockers. A wavelength locker compares the laser output wavelength against a very precise standard. Using a Peltier cooler integral to the laser package, the laser wavelength can then be fine tuned to the precise wavelength required. Tunable lasers require wavelength locking to account for the laser output's drift over time and temperature.

Tunable Filters

Many wavelength lockers use tunable filters (TF), which allow a specific wavelength to be filtered from an optical light wave. Ideally, they function to isolate an arbitrary spectral point at an arbitrary wavelength over a broad spectral range. The very narrow bandwidth of a tunable filter greatly reduces any noise in a DWDM system.

Tunable filters, like Fabry-Perot etalons (discussed in Chapter 9), consist of two highly polished surfaces (mirrors), but these devices also incorporate an area between the two mirrors made up of piezoelectric stacks which determine the separation between the mirrors. This distance sets the bandpass characteristics of the filter. In addition, a highly reflective coating covers the mirror. This coating defines the exact shape and degree of order of separation of the TF's periodic transmission profile.

Figure 7.20: Typical Packaged Tunable Filter

(Photo courtesy of Micron Optics, Inc.)

Any internally reflecting cavity device generates a response given by the periodic Airy function. Figure 7.21 illustrates the airy function response of a typical tunable filter.

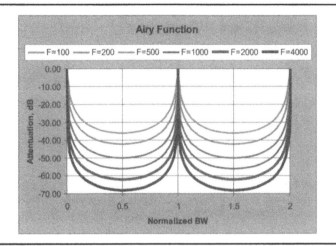

Figure 7.21: Airy Function of a Tunable Filter

(Illustration courtesy of Micron Optics, Inc.)

Chapter Summary

- Semiconductor optical amplifiers (SOA's) are essentially laser diode chips with fiber attached to both ends that amplify and retransmit any incoming optical signal.

- The explosion of dense wavelength-division multiplexing applications have made erbium-doped fiber amplifiers an essential fiber optic system building block.

- The EDFA amplification process uses photons to optically boost the signal; the process is independent of data rate and format, and because the optical output level is so large, fewer amplifiers are required.

- Four applications for EDFA's include: power amplifier, in-line amplifier, preamplifier, and loss compensation.

- Raman amplifiers use stimulated Raman scattering to create optical gain and usually consist of simply a high-power pump laser and a WDM or directional coupler.

- External modulators are named for the fact that the modulation takes place outside of the laser cavity.

- Lithium niobate has become a popular material for external modulators due to its high electro-optic coefficient and ready availability.

- Wavelength converters take an input wavelength and convert it to a different output wavelength using either optoelectronic or all-optical techniques.

- Wavelength lockers compare laser output wavelengths against a precise standard and tune to the laser to a specific wavelength.

- Tunable filters isolate an arbitrary spectral band at an arbitrary wavelength over a broad spectral range.

Selected References and Additional Reading

Chaires, Daryl. "Importance of Mid-Stage Access in Next Generation Optical Amplifiers," from *NFOEC Technical Proceedings, Volume 1*. Proceedings of the National Fiber Optic Engineers Conference, Orlando, FL, Sept. 13-17, 1998.

—. *1713 Erbium-Doped Fiber Amplifier*: Lucent Technologies Technical Note MN97068-57, 1995.

—. *Designer's Guide to External Modulation*. Uniphase Telecommunications Products, 1997.

—. *Using the Lithium Niobate Modulator: Electro-Optical and Mechanical Considerations*: Lucent Technologies Technical Note TN98004, April 1998.

Fuerst, Thomas. "Today's Optical Amplifier — The Cornerstone of Tomorrow's Optical Layer." *Telephony Magazine*, July 7, 1997.

Hiroji Masuda, Shingo Kawai, Ken-Ichi Suzuki, and Kazuo Aida. "Wideband WDM Transmission Using Erbium-doped Fluorite Fiber and Raman Amplifiers." *Optical Society of America, Trends in Optics and Photonics*. TOPS vol. XXV, July 27-29, 1998: 262-265.

Miller, Calvin and Lawrence Pelz. "Fiber Fabry-Perot Tunable Filters Improve Optical Channel Analyzer Performance." www.micronoptics.com. March 1998: 1-3.

Schweber, Bill. "Optical Amplifiers Literally Pump up the (Photon) Volume." *EDN*. July 16, 1998: 40-44.

Y. Sun, J.W. Suihoff, A.K. Srivastava, J.L. Zyskind, T.A. Strasser, J.R. Pedrazzani, C. Wolf, J. Zhou, J.B. Judkins, R.P. Espindola, and A.M. Vengsarkar. "80 nm Ultra-wideband Erbium-doped Silica Fibre Amplifier." *Electronics Letters*, vol. 33, no. 33, November 1997: 1965-1967.

INTERCONNECTION DEVICES

Interconnection devices refer to any mechanism or technique used to join an optical fiber to another fiber or to a fiber optic component. The most common interconnection device is the connector. Connectors were once the most difficult aspect of the commercialization of fiber optics. Today, nothing could be further from the truth. Fiber optic connectors have gone through several development generations in a few years and are now mature and highly reliable devices. The only drawback at this time is the almost bewildering number of connectors to choose from. Table 8.1 summarizes the evolution of fiber optic connectors.

Parameter/Feature	1st Generation	2nd Generation	3rd Generation	4th Generation
Coupling Method	Threaded	Bayonet	Push-Pull	Push-Pull
Ferrule Material	Steel, Brass	Steel, Ceramic	Ceramic, Plastic	Ceramic, Plastic
Alignment Sleeve	Often Loose	Captive	Captive	Captive
Sleeve Material	Plastic	Beryllium, Copper	Beryllium, Copper, Ceramic	Beryllium, Copper, Ceramic
Body Material	Metal	Metal, Plastic	Metal, Plastic	Plastic
Size	Large	Large	Moderate	Small
Rotation Prevention	No	Yes	Yes	Yes
Repeatability	Poor	Good	Very Good	Excellent
Installation Ease	Poor	Good	Very Good	Excellent
Insertion Loss	High	Moderate	Low	Low
Backreflection	Not Addressed	Moderate	Low	Low
Cost	High	Moderate	Low	Low
Multimode Use	Very Good	Very Good	Very Good	Excellent
Single-mode Use	Unusable	Good	Very Good	Excellent
Example	SMA 906	ST^{TM}	SC	LC

Table 8.1: Connector Evolution

OPTICAL CONNECTOR BASICS

Fiber optic connector types are as various as the applications in which they are used. Different connector types have different characteristics, advantages, disadvantages, and performance parameters. But all connectors have the same four basic components.

The Ferrule: The fiber is mounted in a long, thin cylinder, the ferrule, which acts as a fiber alignment mechanism. The ferrule is bored through the center at a diameter that is slightly larger than the diameter of the fiber cladding. The end of the fiber is located at the end of the ferrule. Ferrules are typically made of metal or ceramic, but they may also be constructed of plastic.

The Connector Body: Also called the connector housing, the connector body holds the ferrule. It is usually constructed of metal or plastic and includes one or more assembled pieces which hold the fiber in place. The details of these connector body assemblies vary among connectors, but bonding and/or crimping is commonly used to attach strength members and cable jackets to the connector body. The ferrule extends past the connector body to slip into the coupling device.

The Cable: The cable is attached to the connector body. It acts as the point of entry for the fiber. Typically, a strain-relief boot is added over the junction between the cable and the connector body, providing extra strength to the junction.

The Coupling Device: Most fiber optic connectors do not use the male-female configuration common to electronic connectors. Instead, a dual female coupling device such as an alignment sleeve is used to mate the male connectors. Similar devices may be installed in fiber optic transmitters and receivers to allow these devices to be mated via a connector. These devices are also known as feed-through bulkhead adapters.

Figure 8.1: Parts of a Fiber Optic Connector

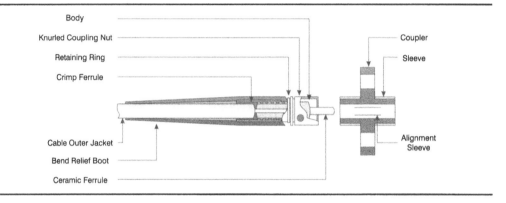

TYPES OF FIBER OPTIC CONNECTORS

There are many fiber optic connector types available, but only a few types are widespread. They are as follows:

SMA 906: SMA 906 connectors represent an old, first generation design now declining in use. They are still used in some military applications because of their ability to withstand high temperatures. Overall, they are difficult to use, generally have poor performance and are suitable only for multimode fiber. They are typically used with 100/140 μm fiber or larger.

ST: ST connectors are very widespread in the U.S. and are used predominately with multimode fiber. The design features a spring loading twist-and-lock bayonet coupling that keeps the fiber and ferrule from rotating during multiple connections. The cylindrical ferrule may be made of plastic, ceramic, or stainless steel. ST connectors offer very good features, cost, and performance. This connector is still popular in military and commercial applications where vibration and shock levels are high. The bayonet coupling is uniquely suited to this environment.

Biconic: The biconic connector was the first highly successful single-mode connector. The ferrule mates to a free-floating alignment sleeve, and a spring between the plug's threaded screw cap and ferrule ensures that a controlled longitudinal force seats the ferrule in the sleeve independent of the amount of tightening. The early models did not address rotation prevention or backreflection, but upgraded designs have addressed these issues. Biconic connectors are widely used by the telecommunications industry and are suitable for single-mode and multimode fiber.

FC: (Also available as FC/PC and FC/APC.) The FC connector features a flat end face on the ferrule that provides "face contact" between joining connectors. The FC represents a very good second generation connector design with very good features and performance but relatively high cost. It offers excellent single-mode and multimode performance and was one of the first connectors to address backreflection. FC's are still widely used for analog systems or high bit-rate systems where backreflection management is important.

FC/PC connectors incorporate a "physical contact" curved polished fiber end face that greatly reduces backreflections. The angled polish connector (APC) version combines a curved end face with an 8° angle to all but eliminate backreflections, even from an unterminated connector.

D4: The D4 connector is keyed, spring loaded, and uses a floating sleeve in its coupler. The ferrule has a diameter of 2.0 mm. It is usable for both single-mode and multimode applications, but at this time, its manufacturer, is its dominant user.

HMS-10: The HMS-10 connectors offer precise construction and fiber alignment. They have low insertion loss, good return loss, and excellent repeatability, but they are very expensive compared to other connectors and are only available completely assembled and terminated to a custom-ordered fiber specification.

SC: The SC (subscription channel) connector uses a locking mechanism that gives an audible click when pushed in or pulled out. This push-pull design prevents rotational misalignment and is intended to be pull-proof, meaning the ferrule is decoupled from the cable and the connector body. A slight pull on the cable will not cause the ferrule to lose optical contact at the interconnection. A duplex version of the SC connector is gaining popularity in networks and other applications requiring full-duplex transmission. The connector is suitable for single-mode and multimode fibers. The SC connector offers excellent packing density as well as exceptional performance and cost. This connector is also popular in an APC configuration.

FDDI: The fiber distributed data interface (FDDI) connector has been in use since 1984. It is used primarily in duplex fiber operations, and it is endorsed by a standard of the same name. This standard describes duplex fiber optic networks. A fixed shroud protects the ferrule from damage. The FDDI connector may also be referred to as a media interface connector (MIC) or a fixed shroud duplex (FSD) connector.

ESCON: Escon connectors, similar to the FDDI connector includes a retractable shroud which pulls back from the ferrule during mating and protects the ferrule when the connector is not in use. Like the FDDI connector, this connector allows the duplex transmission of information.

EC/RACE: This connector uses a rectangular push-pull latching system that makes it nearly impossible to disconnect by pulling on the cable. Originally designed for use in the European community's research and development in advanced communications technologies in Europe (RACE) program, this low-cost, high-density connector is easy to install. High return loss is achieved by incorporating a silicone based index-matching membrane that negates the need for physical contact of the fibers.

LC: The LC connector was developed for premise wiring applications. It uses a body similar to the RJ-45 telephone style housing and a 1.25 mm diameter ferrule that is half the diameter of most other fiber optic connectors.

MT: This is one of many new array connectors for use with ribbon fiber. A single connector can be used with 4, 8, 10, or 12 fibers. The connector itself is little larger than a standard SC connector. It can be used with single-mode or multimode fiber. Single-mode options include several levels of PC polishing (offering different levels of backreflection) as well as APC. This allows for significantly higher packing density compared to all other connector types discussed above.

MT-RJ: This new, dual fiber connector is modeled after the hugely successful RJ-45 electrical connector used in phone systems. It is smaller than an SC connector and provides two fibers. The body style is very similar to the LC connector.

Table 8.2: Connector Types and Characteristics

(**NOTE:** Connector sizes are not scaled and are not relative to one another.)

CONNECTOR	INSERTION LOSS	REPEATABILITY	FIBER TYPE	APPLICATIONS
BICONIC	0.60-1.00 dB	0.20 dB	SM, MM	Telecommunications
D4	0.20-0.50 dB	0.20 dB	SM, MM	Telecommunications
EC/RACE	0.10-0.30 dB	0.10 dB	SM	High-speed Datacom
ESCON	0.20-0.70 dB	0.20 dB	MM	Fiber Optic Networks
FC	0.50-1.00 dB	0.20 dB	SM, MM	Datacom, Telecommunications
FDDI	0.20-0.70 dB	0.20 dB	SM, MM	Fiber Optic Networks
HMS-10	0.10-0.30 dB	0.10 dB	SM	Test Equipment
LC	0.15 dB (SM) 0.10 dB (MM)	0.2 dB	SM, MM	High-density Interconnects
MT ARRAY	0.30-1.00 dB	0.25 dB	SM, MM	High-density Interconnects
SC	0.20-0.45 dB	0.10 dB	SM, MM	Telecommunications
SC DUPLEX	0.2-0.45 dB	0.10 dB	SM, MM	Datacom
SMA	0.40-0.80 dB	0.30 dB	MM	Military
ST	Typ. 0.40 dB (SM) Typ. 0.50 dB (MM)	Typ. 0.40 dB (SM) Typ. 0.20 dB (MM)	SM, MM	Inter-/Intra-building, Security, Navy

The first optical connectors were little more than adaptations of existing electrical connector designs. This yielded functional, although far from optimum, optical connectors. Precision was low and user friendliness was not a consideration. To make matters worse, early fiber was not terribly precise either. The fiber had considerable diameter variation and did not have good concentricity. Because they caused high connector loss, these fiber flaws were most often interpreted by users as poor connector performance, not poor fiber performance. These problems are no longer present in quality modern fibers.

Some examples of early connectors include the SMA 906 types and 38999 types. The SMA 906 connector, one of the earliest widespread types, contains many undesirable features. It is a threaded type, making connection and removal time consuming. Because of the threaded nature of the connector, no two individuals ever tightened the connector the same amount. One person might tighten only to the first resistance, while another might utilize a crescent wrench to grossly over-torque and possibly destroy the connector. Another bad feature of the SMA 906 is the requirement of a plastic alignment sleeve installed between two connectors. If this sleeve is omitted, the connector may be completely unusable or at best, intermittent. This is often called a non-captive alignment sleeve because the alignment sleeve is not permanently built into the connector adapter. The last problem with most first generation connectors is that they allow the connectors to be mated without controlling rotational alignment. This causes a large amount of variation in the insertion loss as the connector is unmated and remated.

To overcome some of the problems cited above, manufacturers developed connectors that used a custom fit jewel to adapt to the varying outer diameter of the fiber. While effective, this approach was very costly and required a great deal of skilled labor for installation. Other unsuccessful approaches included a multitude of expanded beam connectors. Most have been discontinued because an expanded beam connector can never achieve the low loss of a physical contact connector. One clever approach to compensating for variations in fiber dimensions and concentricity is the rotary splice. This splice can be tuned to very low loss levels (<0.05 dB) with almost any fiber. Improved fiber quality decreased the need for such measures.

Fiber optic connectors developed in the first generation were mostly screw-type (threaded) connectors (e.g., SMA 905 and SMA 906). Connectors later evolved to second generation connectors such as the straight tip (ST) type. This connector solved many of the problems associated with the early connectors. Gone is the loose alignment sleeve; it is now captive in the bulkhead adapter. The ST is a bayonet type, assuring that the connector is always properly and consistently connected. With this connector, only one rotational alignment is allowed, greatly improving repeatability. Other notable second generation connectors are the biconic and FC types, although neither address as many first generation flaws as the ST type. The biconic connector is a threaded type and initially did not prevent rotation. Recent design upgrades to the biconic connector have addressed fixed rotational alignment with the keyed biconic. While the FC connector remains popular for single-mode applications, the threaded connector design lacks packing density due to the finger access required to make connections.

Some second generation connectors also incorporated a modified polishing method called PC (physical contact). Earlier connectors used a flat polish on the connector end. This led to a large amount of backreflection, a highly detrimental quality for high bit-rate or analog laser-based fiber optic systems. Large amounts of backreflection in these systems seriously degrade the system's performance and margins. The PC uses a curved polish to dramatically reduce the reflection. Further reduction is achieved by forcing the connector ends together. This eliminates the glass-air-glass interface found in non-PC connectors and further reduces the backreflection. Other variations that achieve low backreflection include angle polishing (angle typically 5° to 15°). Angle polished connectors (APC) require special bulkhead couplings to properly align the connectors. The connectors are then aligned so that the surfaces touch. Names that imply low backreflection include PC, UPC, Super-PC and Super Polish.

Third generation connectors tend to be push-pull types. This makes for the fastest possible connection time and allows a significant increase in packing density because less finger access is required. Third generation connectors have also focused heavily on minimizing installation time (onto the fiber), reducing insertion loss, and addressing backreflection more completely than in earlier generations. Fourth generation connectors are focusing on ease of use and reduced size. Packing density is becoming more important. The LC and MT-RJ connectors are good examples of fourth generation connectors.

INSTALLING FIBER OPTIC CONNECTORS

The method for attaching fiber optic connectors to optical fibers varies among connector types. While not intended to be a definitive guide, the following steps are given as a reference for the basics of optical fiber interconnection.

1. Cut the cable one inch longer than the required finished length.
2. Carefully strip the outer jacket of the fiber with "no nick" fiber strippers. Cut the exposed strength members, and remove the fiber coating. The fiber coating may be removed two ways: by soaking the fiber for two minutes in paint thinner and wiping the fiber clean with a soft, lint-free cloth, or by carefully stripping the fiber with fiber stripper. Be sure to use strippers made specifically for use with fiber rather than metal wire strippers as damage can occur, weakening the fiber.
3. Thoroughly clean the bared fiber with isopropyl alcohol poured onto a soft, lint-free cloth such as Kimwipes®. NEVER clean the fiber with a dry tissue.
4. The connector may be attached by applying epoxy or by crimping. If using epoxy, fill the connector with enough epoxy to allow a small bead of epoxy to form at the tip of the connector. Insert the clean, stripped fiber into the connector. Cure the epoxy according to the instructions provided by the epoxy manufacturer.
5. Anchor the cable strength members to the connector body. This prevents direct stress on the fiber. Slide the back end of the connector into place (where applicable).
6. Prepare the fiber face to achieve a good optical finish by cleaving and polishing the fiber end. Before connection is made, the end of each fiber must have a smooth finish that is free of defects such as hackles, lips, and fractures. These defects, as well as other impurities and dirt change the geometrical propagation patterns of light and cause scattering.

Figure 8.2: Fiber End Face Defects

(Photo courtesy of AMP, Inc.)

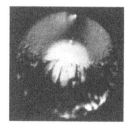 

Cleaving

Cleaving involves cutting the fiber end flush with the end of the ferrule. Cleaving, also called the scribe-and-break method of fiber end face preparation, takes some skill to achieve optimum results. Properly done, the cleave produces a perpendicular, mirror-like finish. Incorrect cleaving will result in lips and hackles as seen in Figure 8.2. While cleaving may be done by hand, a sophisticated cleaver tool, available from such manufacturers as Fujikura, allows for a more consistent finish and greatly reduces the overall skill and time required. The steps listed below outline one procedure for producing good, consistent cleaves such as the one shown in Figure 8.3.

1. Place the blade of the cleaver tool at the tip of the ferrule.
2. Gently score the fiber across the cladding region in one direction. If the scoring is not done lightly, the fiber may break, making it necessary to reterminate the fiber.
3. Pull the excess, cleaved fiber up and away from the ferrule.
4. Carefully dress the nub of the fiber with a piece of 12-micron alumina-oxide paper.
5. Do the final polishing.

The manual cleaving method described above is by far the lowest cost means of cleaving fibers, but it is certainly not the best. Most operators now use automated cleavers to perform this operation. There are some excellent units available in the $1,500 range that require the operator to follow a few simple steps, up to units costing four to six times as much that automate virtually every phase of the cleaving process. In order to achieve low-loss, low-backreflection fusion splices, very high quality cleaves must be achieved. This either requires a highly skilled operator and the manual steps outlined above or a good quality mechanical cleaver.

Figure 8.3: A Well-cleaved Multimode Fiber

(Photo courtesy of AMP, Inc.)

Polishing

Upon achieving a clean cleave, a polishing bushing grinds and polishes the fiber. The proper finish is achieved by rubbing the connectorized fiber end against polishing paper in a figure-eight pattern approximately sixty times. (See Figure 8.4.)

To increase the ease and repeatability of connector installation, most companies offer connector kits. Usually, these kits are specific to the type of connector to be installed while others supply the user with general tools and information for installing different types of connectors.

As mentioned, many connectors require the use of an alignment sleeve. This sleeve increases repeatability between connections. When using SMA connectors, alignment sleeves (see Figure 8.5), also known as half-sleeves or full-sleeves (depending on the length), are required to reduce insertion loss and improve performance.

Figure 8.4: Fiber End Face Polishing Technique

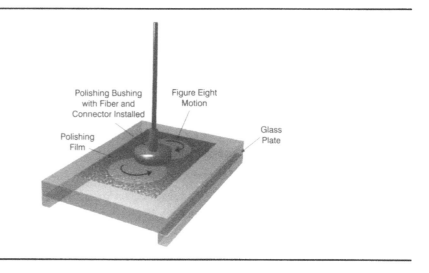

Figure 8.5: Alignment Sleeve

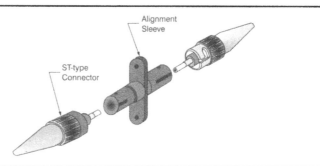

CARE OF FIBER OPTIC CONNECTORS

It is hard to conceive of the size of a fiber optic connector core. Single-mode fibers have cores that are only 8-9 μm in diameter. As a point of reference, a typical human hair is 50-75 μm in diameter, approximately 6-9 times larger! Dust particles can be 20 μm or larger in diameter. Dust particles smaller than 1 μm can be suspended almost indefinitely in the air. A 1 μm dust particle landing on the core of a single-mode fiber can cause up to 1 dB of loss. Larger dust particles (9 μm or larger) can completely obscure the core of a single-mode fiber. Fiber optic connectors need to be cleaned *every* time they are mated; it is essential that fiber optics users develop the necessary discipline to always clean the connectors before they are mated.

It is also important to cover a fiber optic connector when it is not in use. Unprotected connector ends are most often damaged by impact, such as hitting the floor. Most connector manufacturers provide some sort of protection boot. The best protectors cover the entire connector end, but they are generally simple closed-end plastic tubes that fit snugly over the ferrule only. These boots will protect the connector's polished ferrule end from impact damage that might crack or chip the polished surface. Many of the tight fitting plastic tubes contain jelly-like contamination (most likely mold release) that adheres to the sides of the ferrule. A blast of cleaning air or a quick dunk in alcohol will not remove this residue. This jelly-like residue can combine with common dirt to form a sticky mess that causes the connector ferrule to stick in the mating adapter. Often, the stuck ferrule will break off as one attempts to remove it, so always thoroughly clean the connector before mating, even if it was cleaned previously before the protection boot was installed.

Cleaning Technique

Required Equipment:

- Kimwipes or any lens-grade, lint-free tissue. The type sold for eyeglasses work well.
- Denatured alcohol. (Alcohol other than denatured alcohol contains mineral oil and is not suitable for cleaning connectors.)
- 30X microscope.
- Canned dry air.

Technique:

1. Fold the tissue twice, so it is four layers thick.
2. Saturate the tissue with alcohol.
3. First clean the sides of the connector ferrule. Place the connector ferrule in the tissue, and apply pressure to the sides of the ferrule. Rotate the ferrule several times to remove all contamination from the ferrule sides.
4. Now move to a clean part of the tissue. Be sure it is still saturated with alcohol and that it is still four layers thick. Put the tissue against the end of the connector ferrule. Put your fingernail against the tissue so that it is directly over the ferrule. Now scrape the end of the connector until it squeaks. It will sound like a crystal glass that has been rubbed when it is wet.
5. Use the microscope to verify the quality of the cleaning. If it isn't completely clean, repeat the steps with a clean tissue. Repeat until you have a cleaning technique that yields good, reproducible results.
6. Mate the connector immediately! Do not let the connector lie around and collect dust before mating.
7. Air can be used to remove lint or loose dust from the port of a transmitter or receiver to be mated with the connector. Never insert any liquid into the ports.

Handling

1. Never touch the fiber end face of the connector.
2. Connectors not in use should be covered over the ferrule by a plastic protection boot. It is important to note that inside of the protection boots contain a sticky residue that is a by-product of making the dust cap. This residue will remain on the ferrule end after the cap is removed. It is critical to thoroughly clean the ferrule end BEFORE it is mated to the intended unit.
3. The use of index-matching gel, a gelatinous substance that has a refractive index close to that of the optical fiber, is a point of contention between connector manufacturers. Glycerin, available in any drug store, is a low-cost, effective index-matching gel. Using glycerin will reduce connector loss and backreflections, often dramatically. However, the index-matching gel may collect dust or abrasives that can damage the fiber end faces. It may also leak out over time, causing backreflections to increase. The use of index-matching gel is not recommended with modern fiber optic connectors.

SPLICING

Splices are permanent or semi-permanent connections between fibers. Typically, a splice joins lengths of cable outside buildings, while connectors join the ends of cables inside buildings. Splices offer lower attenuation and lower backreflection than connectors, and they are generally less expensive. There are two main types of splices: fusion splices and mechanical splices.

Fusion Splices

Fusion splicing involves butting two cleaved fiber end faces together and heating them until they melt together or fuse. A fusion splicer controls the alignment of the two fibers, keeping losses as low as 0.05 dB. Fusion splicers are relatively expensive devices that usually include an electric arc welder to fuse the fibers, alignment mechanisms, a camera or binocular microscope to magnify the alignment by 50 times or more, and instruments to check the optical power through the fibers both before and after they are fused. The operation of a typical fusion splicer is illustrated in Figure 8.6.

Figure 8.6: Typical Fusion Splicer

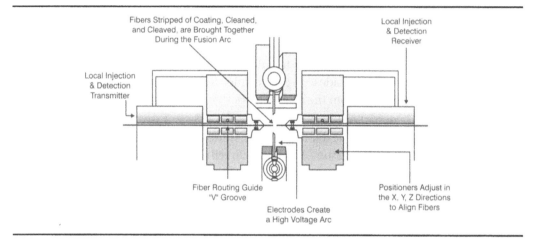

Mechanical Splices

Mechanical splices join two fibers together by clamping them within a structure or by epoxying the fibers together. Generally, the equipment needed to produce a mechanical splice is much less expensive than the equipment for fusion splices. In this case, index-matching gel may be inserted in the mechanical splice to reduce loss, but as stated before, backreflection becomes a concern with this technique.

Capillary splicing is the simplest form of mechanical splicing. It involves inserting two fiber ends into a thin capillary tube as illustrated in Figure 8.7. Index-matching gel is typically used in this splice to keep backreflections at a minimum. Usually, the fibers are held together by compression or friction although epoxy may be used to permanently secure the fibers.

Figure 8.7: Capillary Splice

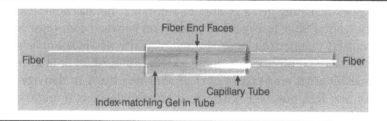

The rotary splice, also called the polished-ferrule splice, involves inserting each clean fiber end into a glass ferrule. After the fiber is secured, the end is polished flush with the ferrule. The two polished ferrules are then butted together in an alignment sleeve that contains index-matching gel. The ferrules have slightly eccentric bores which mount the fibers slightly off-center. After the ferrules are inserted in the alignment sleeve, the ferrules are rotated while the splice loss is monitored. The ferrules are then fixed at the point where splice loss is lowest.

Multiple fiber array splices use grooved top and bottom plates to align multiple fibers with respect to one another. This is especially used to splice ribbon cables. Each fiber in the cable is placed in a separate groove in the bottom plate. When the top and bottom plates are put together, the fibers are automatically aligned.

INTERCONNECTION LOSSES

Connectors and splices both add loss to a system. These losses, like the losses in optical fiber, may be intrinsic or extrinsic. Intrinsic losses occur because of differences in the fibers being connected or spliced. These differences include variations in core and/or outer diameter, differences in the fibers' index profiles, and ellipticity and concentricity of the core. Extrinsic losses are the result of the splice or connector. These loss mechanisms include fiber end misalignment, the quality of the fiber end face finish, contamination on or within the fiber, refractive-index matching between fiber ends, spacing between fiber ends, imperfections in the fiber at the junction, and angular misalignment of the bonded fibers. Chapter 15, "Testing & Measurement Techniques," discusses these loss mechanisms and the methods for measuring these losses in greater detail.

Other Concerns

Fiber curl may also prohibit a successful interconnection two fibers. Fiber curl occurs when the ends of the fiber refuse to lie flat and straight. Although optical fiber curls only a very small degree, even a few micrometers of curl can cause problems in splicing, especially in single-mode fibers. If the two fibers being spliced or connected are not identical, losses can increase. Fiber strength decreases with every splice, because the act of stripping the fiber coating may cause microcracks. In fusion splicing, contaminants can weaken the fusion zone, and thermal cycling can weaken the surrounding area. In connected fibers, the fiber is only as strong as the connectors that join it, but later generation push-pull locking mechanisms can prevent the fibers from being accidentally pulled apart.

FIBER OPTIC CONNECTOR SELECTION GUIDE

Table 8.3 gives general information for selecting fiber optic connectors.

Connector Type	Fiber Type	Generation[1]	Typ. Insertion Loss (dB)[2]	Cost[2] MM	Cost[2] SM
ST	SM, MM	2	0.10-0.25	$1.70-$5.30	$4.35-$11.75
SC	SM, MM	3	0.10-0.34	$2.35-$3.25	$4.50-$7.25
D4	SM	2	0.30	N/A	$10.65
Biconic	SM, MM	2	0.60	$60.00	
SMA	MM	1	0.40-0.60	$4.15-$11.10	N/A
FC	SM, MM	2	0.10-0.25	$4.45-$6.60	$4.95-$8.60
ESCON	MM	3	0.15	$11.60	N/A
FDDI	MM	3	0.30-0.50	$14.90	N/A
LC	SM, MM	4	0.10-0.20	$3.40-$10.50	$5.20-$10.75
MT	SM, MM	4	0.30-1.00	N/A	

Table 8.3: Connector Selection Guide

(Referenced notes are listed after the table.)

Note 1: The Generation column is the estimated generation as described in Table 8.1 at the beginning of the chapter.

Note 2: Source is the 19th edition of *Communication Fiber Optics* catalog published by Fiber Instrument Sales Inc., Oriskany, NY.

Chapter Summary

- Fiber optic connectors have gone through several developmental generations.
- The four basic components of all optical connectors are the ferrule, the connector body, the cable, and the coupling device.
- The method for attaching fiber optic connectors to optical fiber varies from connector type to connector type.
- Cleaving involves cutting the fiber end flush with the end of the ferrule.
- After a cleave has been achieved, the connector is attached to a polishing bushing, and the fiber is ground and polished.
- Most connectors require the use of an alignment sleeve.
- Connectors must always be cleaned before mating, because single-mode fibers have cores that are only 8-9 μm in diameter, and a dust particle anywhere from 9 μm down to 1 μm in diameter can greatly increase optical loss.
- Splices are permanent connections between fibers typically used to join lengths of cable outside buildings.
- Fusion splicing involves butting two cleaved fiber end faces together and heating them until they melt together.
- Mechanical splices join two fibers together by clamping them within a structure or by epoxying the fibers together.
- Different types of mechanical splices include the capillary splice and the AT&T rotary splice.
- Connectors and splices both add loss to a system, while other considerations such as fiber curl and mismatched fibers can further decrease loss.

Selected References and Additional Reading

Ajemian, Ronald G. "A Selection Guide for Fiber Optic Connectors." *Optics & Photonics News.* June 1995: 31-36.

Hecht, Jeff. *Understanding Fiber Optics.* 2nd edition. Indianapolis, IN: Sams Publishing, 1993.

Miller, Mettler, and White. "Optical Fiber Splices and Connectors." NY: Maral Dekken, Inc., 1986.

Reed, Mike. "Making the Perfect Cleave." *Cabling Installation & Maintenance.* April 1995: 38.

Sterling, Donald J. *Amp Technician's Guide to Fiber Optics*, 2nd Edition. New York: Delmar Publishers, 1993.

—. *Universal Transport System Design Guide.* Hickory, NC: Siecor Corporation, 1991.

Weiss, Roger E. "Multifiber-ferrule Ribbon Cable Connector Shrinks Installation Costs." *Lightwave Magazine.* October 1996.

OTHER PASSIVE DEVICES

After the fiber, connectors and splices rank as the most important passive devices in a fiber optic system. Other types of passive devices include couplers, splitters, tap ports, switches, and wavelength-division multiplexers. These devices divide, route, or combine multiple optical signals.

FIBER OPTIC COUPLERS

Fiber optic couplers either split optical signals into multiple paths or combine multiple signals on one path. Optical signals differ from electrical signals, making optical couplers trickier to design than their electrical counterparts. Like electrical currents, a flow of signal carriers, in this case photons, comprise the optical signal. However, an optical signal does not flow through the receiver to the ground. Rather, at the receiver, a detector absorbs the signal flow. Multiple receivers, connected in a series, would receive no signal past the first receiver which would absorb the entire signal. Thus, multiple parallel optical output ports must divide the signal between the ports, reducing its magnitude.

The number of input and output ports, expressed as an N x M configuration, characterizes a coupler. The letter N represents the number of input fibers, and M represents the number of output fibers. Fused couplers can be made in any configuration, but they commonly use even multiples of two (2 x 2, 4 x 4, 8 x 8, etc.).

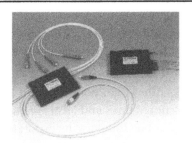

Figure 9.1: Star and Tee Couplers

(Photo courtesy of Hopecom, Inc.)

The simplest couplers are fiber optic splitters. These devices possess at least three ports but may have more than 32 for more complex devices. A simple 3-port device, also called a tee coupler, can be thought of as a directional coupler. One fiber is called the common fiber, while the other two fibers may be called input or output ports. A common application involves injecting light into the common port and splitting it into two independent legs (the output ports). The coupler manufacturer determines the ratio of the distribution of light between the two output legs. Popular splitting ratios include 50%-50%, 90%-10%, 95%-5% and 99%-1%; however, almost any custom value can be achieved. (These values are sometimes specified in dB values.) For example, using a 90%-10% splitter with a 50 μW light source, the outputs would equal 45 μW and 5 μW. However, the outputs never quite do that well due to excess loss. All couplers and splitters share this parameter. Excess loss assures that the total output is never as high as the input. Loss figures range from 0.05 dB to 2 dB for different coupler types. An interesting, and unexpected, property of splitters is that they are symmetrical. For instance, if the same coupler injected 50 μW into the 10% output leg, only 5 μW would reach the common port. Table 9.1 gives the typical insertion losses for modern single-mode couplers.

Table 9.1: Typical
Insertion Losses for
Modern Single-mode
Optical Fibers

Split Ratio (%)	Typical Insertion Loss (dB)
50/50	3.1/3.1
45/55	3.6/2.7
40/60	4.1/2.3
35/65	4.7/2.0
30/70	5.4/1.7
25/75	6.2/1.4
15/85	8.4/0.8
10/90	10.2/0.6
5/95	13.2/0.4
10/45/45	10.5/4.0/4.0
20/40/40	7.3/4.5/4.5
30/35/35	5.4/4.8/4.8
40/30/30	4.1/5.4/5.4
50/25/25	3.1/6.2/6.2
60/20/20	2.3/7.2/7.2
70/15/15	1.7/8.5/8.5
80/10/10	1.0/10.5/10.5
25/25/25/25	6.4/6.4/6.4/6.4

Common applications for couplers and splitters include:

- Local monitoring of a light source output (usually for control purposes).
- Distributing a common signal to several locations simultaneously. An 8-port coupler allows a single transmitter to drive eight receivers.
- Making a linear, tapped fiber optic bus. Here, each splitter would be a 95%-5% device that allows a small portion of the energy to be tapped while the bulk of the energy continues down the main trunk.

In applications that require links other than point-to-point links, optical couplers find the widest use. This includes bidirectional links and local area networks (LAN). In LAN applications, either a star topology or a bus topology incorporate couplers. In a star topology, stations branch off from a central hub, much like the spokes on a wheel. The allows easy expansion of the number of workstations; changing from a 4 x 4 to an 8 x 8 doubles the system capacity. The star coupler divides all outputs allowing every station to hear every other station. Star couplers have many ports (usually a power of two), and couplers with 32 or 64 ports are not uncommon. One use of a star coupler creates a large party-line circuit. Many transceivers connect to the star coupler and can communicate with all other transceivers, assuming the network adopts a protocol which prevent two or more transceivers from communicating simultaneously. Large insertion loss, (20 dB typically for a 64-port device) creates the biggest disadvantage of the star coupler, as is the need for a complex collision-prevention protocol.

Bus topology utilizes a tee coupler to connect a series of stations that listen to a single backbone of cable. In a typical bus network, a coupler at each node splits off part of the power from the bus and carries it to a transceiver in the attached equipment. In a system with N terminals, a signal must pass through N-1 couplers before arriving at the receiver. Loss increases linearly as N increases. A bus topology may operate in a single direction or a bidirectional or duplex configuration. In a one way, unidirectional setup, a transmitter at one end of the bus communicates with a receiver at the other end. Each terminal also contains a receiver. Duplex networks add a second fiber bus or use an additional directional coupler at each end and at each terminal. In this way, signals flow in both directions. Figure 9.2 illustrates the differences in a unidirectional bus and a duplex bus.

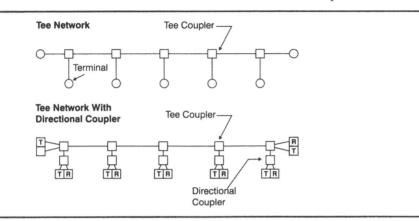

Figure 9.2: Tee Network Configurations

WAVELENGTH-DIVISION MULTIPLEXING (WDM)

The fiber optic industry first deployed single wavelength transmission links. As requirements changed, the industry responded with wavelength-division multiplexing (WDM), which sends two distinct signals per fiber, doubling transmission capacity. Like the simple splitter, WDM's typically have a common leg and a number of input or output legs. Unlike the splitter, however, they have very little insertion loss. They do have the same range of excess loss. Two important considerations in a WDM device are crosstalk and channel separation. Crosstalk, also called directivity, refers to separation of demultiplexed channels. Each channel should appear only at its intended port and not at any other output port. The crosstalk specification expresses how well a coupler maintains this port-to-port separation. Channel separation describes a coupler's ability to distinguish wavelengths. In most couplers, the wavelengths must be widely separated. This allows light to travel in either direction without the penalty found in splitters. WDM's allow multiple independent data streams to be sent over one fiber. The most common WDM system uses two wavelengths, although four or more-wavelength systems are available.

Figure 9.3 shows two WDM's allowing bidirectional streams of data to be carried on a single fiber. The type of data streams does not matter. One could be a video signal, the other an RS-232 data stream. Alternatively, both signals could be video signals or high-speed data signals at 2.488 Gb/s.

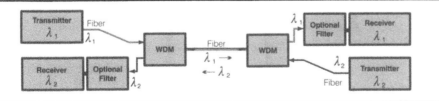

Figure 9.3: Bidirectional WDM Application

Figure 9.4 illustrates a bulk optics WDM. Constructed from discrete lenses and filters, a dichroic filter lies at the heart of this type of WDM. Dichroic filters, based on interferometric techniques, reflect the light that they do not transmit. Referring to the figure, imagine that Fiber 1 carries two wavelengths, 850 nm and 1310 nm. Also imagine that the dichroic filter passes wavelengths longer than 1100 nm, known as long-wave pass (LWP) filter. As the light exits Fiber 1 it first passes through the lens which focuses the light at a point. As the light hits the filter, the 1310 nm light passes through the filter and is collected by Fiber 3. The 850 nm light exiting Fiber 1 on the other hand reflects off of the filter and is collected by Fiber 2. Thus the two wavelengths have been effectively separated, and the information carried on each wavelength can be independently decoded. The dichroic filter can offer a great deal of isolation in the transmission mode, but has poor isolation in the reflection mode. Usually these types of WDM's feature both short-wave pass (SWP) and LWP filters, and these filters combine to achieve the best system performance.

Figure 9.4: Bulk Optics WDM

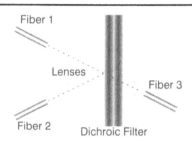

COARSE WAVELENGTH-DIVISION MULTIPLEXING (CWDM)

In response to the ever-growing fiber network demand, the fiber optic industry developed an intermediate technology called CWDM (coarse wavelength-division multiplexing). With a capacity greater than WDM and smaller than DWDM, discussed next, CWDM allows a modest number of channels, typically eight or less, to be stacked in the 1550 nm region of the fiber called the C-Band. In order to dramatically reduce cost, CWDM's use uncooled lasers with a relaxed tolerance of ± 3 nm. Whereas a DWDM system has channels spaced as close as 0.4 nm, CWDM uses a spacing of 20 nm. The wide spacing accommodates the uncooled laser wavelength drift that occurs as the ambient temperature varies. The wavelength of uncooled lasers drifts about ± 0.06 nm/°C. CWDM transmission may occur at one of eight wavelengths: typically 1470 nm, 1490 nm, 1510 nm, 1530 nm, 1550 nm, 1570 nm, 1590 nm, and 1610 nm.

Figure 9.5: CWDM Passband Characteristics for an 8 Channel Device

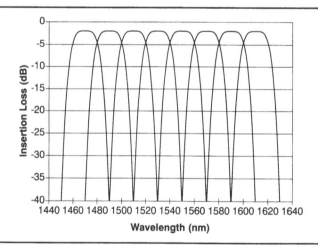

Figure 9.5 shows the typical passband characteristics of a eight channel CWDM multiplexer. Insertion loss for an eight channel device is about 2 dB per end. The passband is about 13 nm wide at the -0.5 dB loss point. CWDM demultiplexers typically have higher insertion loss and significantly better isolation loss. The multiplexers have lower insertion loss and poorer isolation loss. When used in a unidirectional application, the mutiplexer combines various transmitter outputs, so isolation does not matter. In bidirectional appli-

cations, any input on either end of the fiber can be an input or an output, requiring the higher isolation of demultiplexers to guarantee that the system will work without interference between channels.

DENSE WAVELENGTH-DIVISION MULTIPLEXING (DWDM)

Dense wavelength-division multiplexing involves sending a large number of closely spaced optical signals over a single fiber. Standards developed by the ITU (International Telecommunications Union) define the exact optical wavelengths used for DWDM applications. The center of the DWDM band at lies 193.1 THz with standard channel spacings of 200 GHz and 100 GHz. The closest "standard" spacing (100 GHz) allows transmission of 45 channels on one fiber. A 45 channel system spaced at 100 GHz would cover a optical span of 35 nm and require a costly wide bandwidth, gain-flattened EDFA.

As system designers looked to pack more than the 45 channels at 100 GHz spacing, they started to use closer spaced optical channels. The channel spacing, in GHz, relates to the optical wavelength as follows: A spacing of 200 GHz corresponds to about 1.6 nm, 100 GHz corresponds to about 0.8 nm, and 50 GHz corresponds to about 0.4 nm channel spacing. Most commonly 50 GHz follows 100 GHz, although attempts at 75 GHz and 37.5 GHz show up in literature. While there is nothing magical about any of these numbers, it seems likely that 50 GHz will be the next logical step below 100 GHz. Using a channel spacing of 50 GHz (0.4 nm) allows 45 channels to occupy only 17.5 nm of optical bandwidth. This greatly simplifies the requirement for optical amplifiers in the system. Further increases in channels per fiber would likely lead to the use of 25 GHz spacing.

Designing the optical demultiplexer to separate the signals at the receive end defines the greatest challenge in closely spaced optical channels. Because of subtle color differences in each of the optical channels, high performance DWDM optical demultiplexers must have three characteristics. First it must be very stable over time and temperature. Second, it needs to have a relatively flat passband or region of frequencies. Third, it must reject adjacent optical channels so that they do not interfere. Several basic types of designs can be used in optical demultiplexers to separate the optical channels. Many of these designs have an increasingly difficult time separating the optical channels as the spacing becomes very close. Some, however, such as fiber Bragg gratings (FBG), discussed later in this chapter, actually appear better suited for closer channel spacing. The need for close optical channel spacing is a trade-off between the performance required of the optical amplifiers used in the system and the number of channels to be transmitted per fiber. Figure 9.6 illustrates the transmission spectra of 0.4 nm spacing DWDM FBG's.

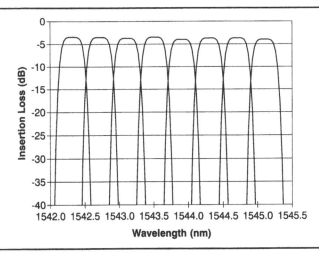

Figure 9.6: 0.4 nm Channel Spacing DWDM FBG's

Red and Blue Bands

The ITU approved DWDM band extends from 1528.77 nm to 1563.86 nm, and divides into the red band and the blue band. The red band encompasses the longer wavelengths of 1546.12 nm and higher. The blue band wavelengths fall below 1546.12 nm. This division has a practical value because useful gain region of the lowest cost EDFA's corresponds to the red band wavelengths. Thus if a system only requires a limited number of DWDM wavelengths using the red band wavelengths yields the lowest overall system cost.

ARRAYED WAVEGUIDE GRATING (AWG)

Arrayed waveguide gratings (AWG's), which incorporate planar lightwave circuits (PLC), have emerged to further the DWDM boom. Constructed primarily from silica on silicon, high performance PLC's require the same tools and techniques as IC's for fabrication. These lithographic techniques build various layers and features on the silicon substrate.

Figure 9.7: Cross-section of a PLC

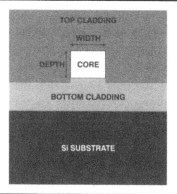

Figure 9.7 shows the cross-section of a PLC. The key features of an optical fiber, the core and cladding, can be seen. In this case, the core is rectangular, not round. Based on interference principles, an AWG device, consists of an array of curved-channel waveguides with fixed differences in the path lengths between adjacent waveguides.

Figure 9.8 illustrates a typical AWG. The waveguides connect to lens regions at the input and output. Light enters the input cavity, splitting the optical power into numerous waveguides, often more than 100. Each waveguide has a slightly different length from the other waveguides. The optical length difference of each waveguide introduces varying delays. The multiple waveguides all terminate into a lens/mixing region. The mixing process results in different wavelengths having maximal interference at different locations, which correspond to the output ports.

Figure 9.8: Operation of an AWG

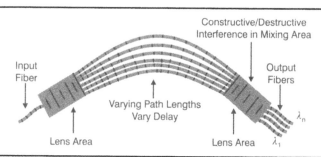

FUSED FIBER COUPLERS

By far the most popular type of coupler in use today is a fused fiber coupler. In this type of coupler, two or more fibers are twisted together and melted in a flame. Figure 9.9 shows the basic construction.

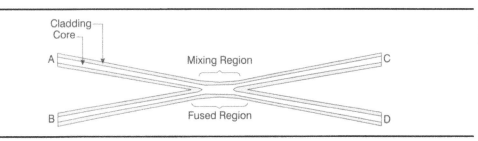

Figure 9.9: Fused Fiber Coupler

This construction technique can be used to make 50%-50% couplers, 99%-1% couplers and even WDM's. The length of the coupling region (the fused region) as well as the amount of twisting and pulling done on the fiber while it is melted, determines the result. This coupler has grown in popularity because of the low cost of the basic materials needed for its construction: a few meters of fiber, a bit of potting compound, and a metal tube. The magic is knowing how to melt, twist, and pull the fiber.

The most interesting type of fused fiber coupler is the WDM. It is only possible with single-mode fiber. An interferometric action forms the WDM within the fused mixing region. Like an interferometer, this causes a sinusoidal response as the length increases. WDM's operate at two specific wavelengths. Adjusting the minimum of the sinusoid to correspond to the first wavelength of interest and the maximum of the sinusoid to correspond to the second wavelength of interest forms a WDM.

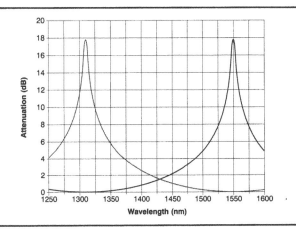

Figure 9.10: Optical Characteristics of a Fused Fiber WDM

Figure 9.10 shows the optical characteristics of a fused fiber coupler WDM that has been designed for wavelengths of 1310 nm and 1550 nm. Port 1 (gray curve) would be designated for use with 1310 nm while Port 2 (black curve) would be designated for use with 1550 nm. While the shapes of the curves may appear unfamiliar, noted that they form by taking the logarithm of a raised sine function.

The WDM shown in the figure has a maximum isolation of about 18 dB. That may be adequate for some applications, but many will require even greater isolation. Cascading two or more WDM's increases the isolation. Usually this type of WDM only uses laser sources because of the relatively narrow isolation peak at each wavelength. However, the passband peak is quite broad.

FIBER OPTIC SWITCHES

Many optical networks incorporate optical switches. Networks that require protection switching (switching between redundant paths), where key attributes must be reliable operation after a long period in one position, system monitoring, and diagnosis commonly feature these devices. Speed is not a crucial parameter for these applications, as speeds as high as tens of milliseconds are acceptable. Dynamic optical routing in the future will require much faster switching speeds. Figure 9.11 illustrates typical switch configurations

Figure 9.11: Typical Switch Configurations

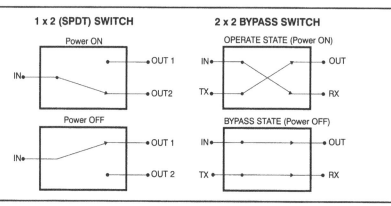

More technologies exist for optical switching than any other function within the optical network. Four main types of optical switches include opto-mechanical switches, thermo-optic switches, electro-optic switches, and all optical switches.

Opto-mechanical Switches

These are the oldest type of optical switches and the most widely deployed at this time. These devices achieve switching by moving fiber or other bulk optic elements by means of stepper motors or relay arms. This causes them to be relatively slow with switching times in the 10-100 ms range. They can achieve excellent reliability, insertion loss, and crosstalk. Usually, opto-mechanical optical switches collimate the optical beam from each input and output fiber and move these collimated beams around inside the device. This allows for low optical loss, and allows distance between the input and output fibers without deleterious effects. These devices have more bulk compared to other alternatives, although new micro-mechanical devices overcome this. Figure 9.12 shows an opto-mechanical switch.

Figure 9.12: FDDI Optical Switch

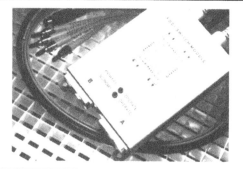

Thermo-optic Switches

These are usually based on waveguides made in polymers or silica. For operation, they rely on the change of refractive index with temperature created by a resistive heater placed above the waveguide. Their slowness does not limit them in current applications.

Electro-optic Switches

These are typically semiconductor-based (e.g., InP), and their operation depends on the change of refractive index with electric field. This characteristic makes them intrinsically high-speed devices with low power consumption. Neither the electro-optic or thermo-optic optical switches can yet match the insertion loss, backreflection and long-term stability of opto-mechanical optical switches.

CIRCULATORS

Circulators such as the one illustrated in Figure 9.13, are three-port devices that have the following property; Port 1 couples to Port 2 and Port 2 couples to Port 3. No other selection of ports conducts light. Light will not even travel in the reverse directions, Port 2 to Port 1 and Port 3 to Port 2. Circulators are often used in conjunction with Bragg gratings (discussed next), since Bragg gratings usually work in a reflective mode. The Bragg grating can be placed at Port 2 of the circulator. Light that is input to Port 1 travels out of Port 2, reflects off of the Bragg grating and then reenters Port 2 where it then exits Port 3. Circulators are also used with lengths of dispersion-compensating fiber (DCF). The DCF is placed at Port 2 with a reflector at the end of the DCF. The light entering Port 1 of the circulator travels through the DCF twice before exiting Port 3 of the circulator. This way the user gets twice as much benefit from a given length of DCF.

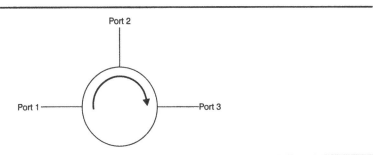

Figure 9.13: Circulator

BRAGG GRATINGS

In-fiber Bragg gratings are simple reflecting elements "written" into the core of optical fibers. This is usually accomplished by treating the fiber so that it becomes photosensitive and then exposing the fiber to UV light through a grating. This introduces periodic variations in the refractive index of the fiber. As light waves interact with the variations in the refractive index, interference effects occur which cause certain wavelengths to reflect more than others. The size and spacing of the reflective elements allows the creation of an almost infinite number of filter responses. Consequently, the design possibilities offer a variety of unique WDM devices based in fiber Bragg gratings.

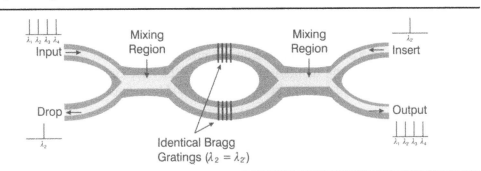

Figure 9.14: Bragg Gratings in a Mach-Zehnder Configuration

(Illustration courtesy of Highwave Optical Technologies.)

Figure 9.14 shows two identical Bragg gratings formed in the legs of a Mach-Zehnder interferometer. The Bragg gratings are designed to reflect a wavelength λ_2. Four discrete wavelengths are applied to the input leg. Wavelength λ_2 reflects off of both Bragg gratings and exits at the "Drop" leg. A new signal at wavelength $\lambda_{2'}$ ($=\lambda_2$) inputs at the "Insert" leg. It reflects off of the Bragg gratings and exits along with λ_1, λ_3 and λ_4. This function, common in DWDM systems, is called a "drop and insert" function.

Bragg gratings are essentially narrow-band devices but can be made broad-band by a variety of means, e.g. by incorporating chirp (non-uniform spacing of the reflecting elements). Functionality can be increased easily, with sub-dB insertion loss, by combining various types of fiber gratings, narrow or broad bandwidth, with or without designed chirp, along with other optical components.

Figure 9.15: Fiber Bragg Grating Used to Compensate for Chromatic Dispersion

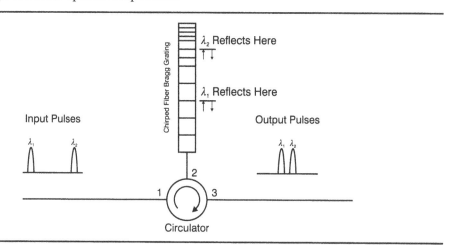

Another very important use for Bragg gratings is compensating for chromatic dispersion. As discussed earlier, chromatic dispersion occurs when the different wavelengths of light in a pulse travel at different speeds, thus arriving at different times. Bragg gratings offer a simple way of correcting this problem. Consider a system containing two distinct wavelengths of light that have become separated in time. Figure 9.15 shows a fiber carrying two wavelengths, λ_1 and λ_2. In this case, λ_2 is ahead of λ_1. As the pulses reach the circulator, they enter port 1 and exit port 2. Both λ_1 and λ_2 pulses travel vertically in the chirped fiber Bragg grating. At about the middle of the length, λ_1 encounters a region of disturbances of refractive index that causes it to reflect and travel back down the grating. Wavelength λ_2 travels further up the grating before it reflects. When both λ_1 and λ_2 pulses reach the circulator, they reenter port 2 and are directed to port 3. Since λ_2 traveled farther than λ_1, the time spacing between the two pulses at the output reduces compared to the input. The FBG adjusts so that the two output pulses overlapped. This basic principle compensates for chromatic dispersion in long-haul fiber optic transmission systems.

Fiber Bragg gratings provide an economical and effective means of canceling out dispersion due to the optical fiber as well as gain equalization throughout the system required primarily to correct for the non-flat response of the EDFA's. Figure 9.16 shows various applications of fiber Bragg gratings in a modern DWDM long-haul fiber system.

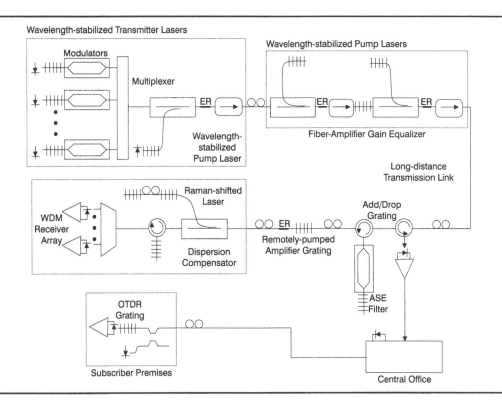

Figure 9.16: High-performance DWDM System Using Fiber Bragg Gratings

ETALON

Etalons, also know as Fabry-Perot (FP) etalons, inserted in a laser cavity, use mirrored surfaces within the laser cavity to achieve a narrowing of the wavelength spread of the emitted laser light. Fabry-Perot etalons, put simply, are Fabry-Perot interferometers with a fixed distance between the two mirrors of the interferometer. Typical FP interferometers incorporate two thick glass plates to enclose a parallel section of air. The inner surfaces of the glass plates are coated to a mirror finish. The outer surfaces tilt at a slight angle relative to the inner surfaces to eliminate spurious interference patterns that may occur from the beams reflecting on the mirror sides. After the light is reflected between the two mirrors multiple times, a converging lens is used to concentrate the light on a screen. In an FP interferometer, the distance between the mirrors may vary depending on the application. With an FP etalon, the mirror distance is fixed, allowing the FP etalon to produce a very narrow laser output. Figure 9.17 illustrates an example of an FP etalon inserted into a laser cavity.

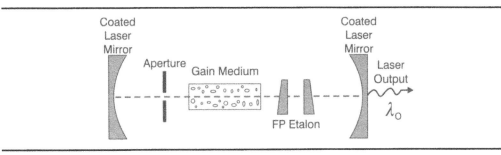

Figure 9.17: FP Etalon in a Laser Cavity

Chapter Summary

- Passive devices include couplers, splitters, tap ports, switches, and WDM's.

- Common applications for couplers and splitters include local monitoring of a light source, distributing a common signal to several locations simultaneously, and making a linear, tapped, fiber optic bus.

- WDM's allow two or more different wavelengths of light to be split into multiple fibers or combined onto one fiber.

- CWDM's allow transmission of up to eight signals in the wavelength range of 1470 nm to 1610 nm.

- DWDM's send a large number of closely spaced optical signals over a single fiber. These signals range in wavelength from 1528.77 nm to 1563.86 nm, encompassing the red and blue bands of the electromagnetic spectrum.

- AWG's incorporate planar lightwave circuits that allow the transmission of multiple wavelengths.

- Fiber optic switches are components that selectively route optical signals among different fiber paths without optical-to-electrical conversion. Three main types of switches include opto-mechanical, thermo-optic, and electro-optic.

- Circulators are one-way, three-port devices, used in fiber Bragg gratings, where light at port one couples to port two, and light at port two couples to port three.

- Bragg gratings, simple reflecting elements in the fiber's core, cause certain wavelengths of light to reflect more than others.

- An etalon is simply a Fabry-Perot interferometer with a fixed distance between the two mirrors in interferometer.

Selected References and Additional Reading

Benech, Elodie. "Fiber Bragg Gratings — Use and Application." Highwave Optical Technologies, Letter to Force, Inc. December 9, 1998.

Cheng, Yale and Gary Duck. "High-Performance Optical Circulators for Fiber Optic Communication Applications." Paper presented at the National Fiber Optic Engineers Conference. June 18-22, 1995. Boston, MA: 971-976.

Dong, L., M.J. Cole, D. Ellis, M. Durkin, M. Ibsen, V. Gusmeroli, and R.I. Laming. "40 Gbit/s 1.55 micron Transmission Over 109 km of Non Dispersion-shifted Fibre with Long Continuously Chirped Fibre Gratings." *OFC Conference Papers.* Volume 6. February 1997: PD6-1 - PD6-4.

—. "Mirrors and Etalons." The Center for Occupational Research and Development, Module 6-5. Waco, Texas: 1987, 1-30.

Palladino, John R. *Fiber Optics: Communicating By Light.* Piscataway, NJ: Bellcore, 1990.

Pan, J.J and Y. Shi. "Dense WDM Multiplexer and Demultiplexer with 0.4 nm Channel Spacing." *Electronics Letters.* Volume 34, No. 1. January 8, 1998: 74-75.

Pesavento, Gerry, and Dr. David Polinksy. "Fiberoptic Switches." *Fiber Optic Product News.* Feb. 1993.

Sterling, Donald J. *Amp Technician's Guide to Fiber Optics,* 2nd Edition. New York: Delmar Publishers, 1993.

—. *Universal Transport System Design Guide.* Hickory, NC: Siecor Corporation, 1991.

Willer, Alan, Kai-Ming Feng, Jin-Xing Cai, Jack Feinberg, Dmitry Starodubov, and Victor Grubsky. "Fiber Grating Varies Dispersion Compensation in Real Time." *Photonics Online.* March 19, 1998.

SYSTEM DESIGN CONSIDERATIONS

10

SYSTEM DESIGN FACTORS

To achieve high-quality transmission, careful decisions based on operating parameters apply for each component of a fiber optic transmission system. The main questions, given in Table 10.1, involve data rates and bit error rates in digital systems, bandwidth, linearity, and signal-to-noise ratios in analog systems, and in all systems, transmission distances. These questions of how far, how good, and how fast define the basic application constraints. Once these are decided, it is time to evaluate the other factors involved.

System Factor	Consideration/Choices
Type of Fiber	Single-mode or Multimode
Dispersion	Regenerators or Dispersion Compensation
Fiber Nonlinearities	Fiber Characteristics, Wavelengths, and Transmitter Power
Operating Wavelength	780, 850, 1310, 1550 nm, and 1625 nm typical.
Transmitter Power	Typically expressed in dBm.
Source Type	LED or Laser
Receiver Sensitivity/Overload Characteristics	Typically expressed in dBm.
Detector Type	PIN Diode, APD, or IDP
Modulation Code	AM, FM, PCM, or Digital
Bit Error Rate (BER) (Digital Systems Only)	10^{-9}, 10^{-12} Typical
Signal-to-Noise Ratio	Specified in decibels (dB).
Number of Connectors	Loss increases with the number of connectors.
Number of Splices	Loss increases with the number of splices.
Environmental Requirements	Humidity, Temperature, Exposure to Sunlight
Mechanical Requirements	Flammability, Indoor/Outdoor Application

Table 10.1: Factors for Evaluating Fiber Optic System Design

Many of these considerations are directly related to other considerations. For example, detector choice will impact the receiver sensitivity which will affect the necessary transmitter output power. Output power impacts the transmitter light emitter type which will affect the usable fiber type and connector type. A logical way to proceed with designing a fiber link involves analyzing the fiber optic link power budget or optical link loss budget.

OPTICAL LINK LOSS BUDGET

Figure 10.1 shows a simple means of visualizing the key required optical calculations for designing a fiber optic link. A practical link must tolerate some range of optical loss. Ideally, but not always, it should work back-to-back (i.e., with the shortest possible fiber). And of course, it should work with some longer length of fiber. Factors affect transmission distance over optical fiber include:

- Transmitter optical output power
- Operating wavelength
- Fiber attenuation
- Fiber bandwidth
- Receiver optical sensitivity

The designer can often adjust any or all of these variables to create a product that meets the needs of a given application.

Figure 10.1: Optical Link Loss Budget

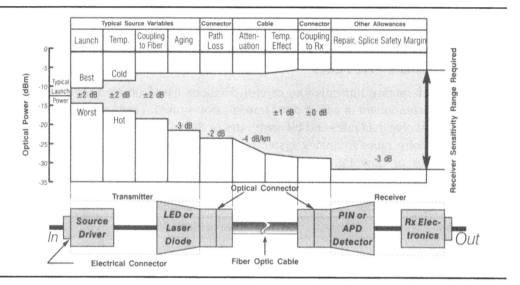

The graphic shows a hypothetical link and its corresponding link budget. Start with the transmitter output power on the left side of the chart. The typical launch power is -12.5 dBm. However, the transmitter LED output power can vary by ±2 dB due to manufacturing variability of the LED itself. Therefore, the output power can be as high as -10.5 dBm or as low as -14.5 dBm. The block is shaded between these two values.

Further transmitter variations of ±2 dB result from the effects of temperature on the electronics and the electro-optics (e.g., LED or laser). Another potential ±2 dB of loss is due to variations in the optical coupling to the transmitter output. The effects of aging, typically 1-3 dB, should be included in the system's design. The next factor involves the losses due to optical connectors that may be in the optical path. The graphic allows 2 dB for this factor. For this system, the loss due to the optical fiber itself amounts to 4 dB/km of length. Multiply this value times the actual length to determine the loss due to the fiber. considerations for temperature effects associated with most fibers usually yield ±1 dB.

The next factor, variation in loss at the receiver, requires a large-area detector to eliminate the effects of this parameter. Finally, a 3 dB safety margin should be built into all systems. At each step, any variation causes the shaded band to enlarge. On the right side of the chart the receiver has to cope with optical inputs as high as -5.5 dBm and as low as -31.5 dBm. Or stated differently, the receiver would need an optical loss range or optical dynamic range of 26 dB.

A discussion of the decibel is necessary to understand these link loss values. The decibel (dB) is a convenient means of comparing two powers. The loss a given link can tolerate is rated in dB. For example, a given AM video link may tolerate a maximum of 9 dB of optical loss. How much light actually reaches the receiver? Table 10.2 describes the decibel to power conversion. According to the Table 10.2, 12% of the optical power actually reaches the receiver, so 88% of the light output by the transmitter was lost somewhere along the way. If the link could tolerate 20 dB of optical loss, only 1% of the transmitter's optical output would reach the receiver. To determine the amount of light reaching the receiver, take any two values that total the dB of optical loss in question. For example, 15 dB is the total of 10 dB and 5 dB. The corresponding power out for 15 dB is 3.2% according to Table

10.2. This value is also attainable by multiplying the corresponding percent values for the two dB readings, 10 dB and 5 dB, to get the desired result, e.g. 10% times 32% is 3.2%. Thus, 3.2% of the light actually reaches the receiver.

dB	Power Out as a % of Power In	% of Power Lost	Remarks
1	79%	21%	...
2	63%	37%	...
3	50%	50%	1/2 the power
4	40%	60%	...
5	32%	68%	...
6	25%	75%	1/4 the power
7	20%	80%	1/5 the power
8	16%	84%	1/6 the power
9	12%	88%	1/8 the power
10	10%	90%	1/10 the power
11	8.0%	92%	1/12 the power
12	6.3%	93.7%	1/16 the power
13	5.0%	95%	1/20 the power
14	4.0%	96.0%	1/25 the power
15	3.2%	96.8%	1/30 the power
16	2.5%	97.5%	1/40 the power
17	2.0%	98.0%	1/50 the power
18	1.6%	98.4%	1/60 the power
19	1.3%	98.7%	1/80 the power
20	1.0%	99.0%	1/100 the power
25	0.3%	99.7%	1/300 the power
30	0.1%	99.9%	1/1000 the power
40	0.01%	99.99%	1/10,000 the power
50	0.001%	99.999%	1/100,000 the power

Table 10.2: Decibel to Power Conversion

A decibel is always a ratio between two numbers. Equation 10.1 gives the calculation of a decibel.

Eq. 10.1
$$dB = 10 \cdot \log_{10}\left(\frac{P_1}{P_2}\right)$$

For fiber optics, the ratio is generally the transmitter output power compared to the receiver input power as in the examples above. The decibel describes all loss mechanisms in the optical path of a fiber optic link. This includes fiber loss (usually described as decibels per kilometer or dB/km), connector loss, and splice loss. In all cases, Table 10.2 provides easy reference for converting the decibel value to a percentage. There is also a unit of power called dBm which indicates the actual power level referred to 1 milliwatt. The conversion table is useful here as well. In the case of the dBm, the ratio is between some power level to be described and 1 milliwatt. Typically the number is negative because power levels are typically less than 1 milliwatt in fiber optics. Table 10.3 shows some typical dBm values, converts them to power, and relates them to actual applications.

Table 10.3: Typical dBm Values

dBm Value	% of 1 Milliwatt	Power	Application
0.0	100%	1.0 milliwatt	Typical Laser Peak Output
-13.0	5%	50.0 microwatts	Typical LED Peak Output
-30.0	0.1%	1.0 microwatt	Typical PIN Receiver Sensitivity
-40.0	0.01%	100.0 nanowatts	Typical APD Receiver Sensitivity

Newcomers to purchasing fiber optics often mistakenly feel they must over-specify the system to include transmitter power, receiver sensitivity, *and* optical link loss budget. In most cases, the customer needs to specify only the optical link loss budget. A customer needing a 10 dB maximum optical link loss budget will never know the difference between a system with a transmitter with a 0 dBm output and a receiver with -10 dBm sensitivity and a system with a transmitter with a -10 dBm output and a receiver with -20 dBm sensitivity. Meeting all other requirements, such as bit error rate (BER), completes the application's loss requirements. By specifying only the required maximum optical loss, the most economical transmitter/receiver pair can be utilized. The only time that transmitter optical output power and receiver optical sensitivity need to be specified is when the transmitter and receiver are bought separately. In that case, the maximum optical link loss budget need not be specified.

RISE-TIME BUDGET

The optical link loss budget analyzes the link to ensure that sufficient power is available throughout the link to meet the demands of a given application, but this power is only one part of the link requirement. The other part is bandwidth or rise time. To meet bandwidth requirements the components in the link must turn on and off quickly, and fiber dispersion must be kept to a minimum. It is easy to overlook bandwidth in many fiber optic applications; the current media hype constantly emphasizes fiber's unlimited bandwidth, but this isn't always the case. Adequate bandwidth for a system can be assured by developing a rise-time budget. As noted in the previous chapters on light emitters and light detectors, the devices do not turn on or turn off instantaneously. Rise and fall times determine the overall response time and the resulting bandwidth. In Chapter 3, we learned that dispersion also limits bandwidth. When the bandwidth of a component is given, the 10% to 90% rise time can be derived from:

Eq. 10.2
$$t_r = \frac{0.35}{BW}$$

Where:

t_r = Rise time.
BW = Bandwidth.

This equation accounts for multimode dispersion in the fiber. The rise time budget must also include the rise times of the transmitter and the receiver. For the receiver, rise time/bandwidth may be limited by either the rise time of the components or the bandwidth of the resistive-capacitive (RC) time constant. Because connectors, couplers, and splices do not affect system speed, they need not be accounted for in the rise time budget as they were in the optical link loss budget. Once all the necessary component rise times have been determined, the system rise time can be derived from:

Eq. 10.3
$$t_{sys} = 1.1\sqrt{t_{r1}^2 + t_{r2}^2 + ... + t_m^2}$$

The 1.1 factor at the beginning of the equation allows for a 10% degradation factor in the system rise time. The system rise-time budget can set any component rise time simply by rearranging the equation to solve for the unknown rise time. Table 10.4 gives an example of a rise-time analysis for a simple fiber optic system.

Element	Bandwidth	Rise Time (10%-90%)
Tx Drive Electronics	200 MHz	1.75 ns
LED (850 nm)	100 MHz	3.50 ns
Fiber (1 km)	90 MHz	3.89 ns
PIN Detector	350 MHz	1.00 ns
Rx Electronics	180 MHz	1.94 ns

Table 10.4: Rise-time Analysis

The system rise time would be found by applying Equation 10.3 to the individual element rise times listed above. Equation 10.3 yields a system rise time of 6.53 ns. Equation 10.2 can then be used to determine that the system bandwidth is 53.6 MHz. The slowest element in the system dominates in setting the overall system bandwidth. For example, using the system given in Table 10.4, changing the LED to a more expensive 200 MHz type would only improve the system's bandwidth by 16% to 62.4 MHz because of the effect of the low fiber bandwidth.

SENSITIVITY ANALYSIS

A sensitivity analysis determines the minimum optical power that must be present at the receiver in order to achieve the performance levels required for a given system. Several factors will affect this analysis:

Source Intensity Noise: This refers to noise generated by the LED or laser; there are two main types. Phase noise describes the difference in the phases of two optical wavetrains, separated by time, cut out of the same optical wave. When a comparison of the two trains shows a stable phase difference, the wave is coherent. An unstable phase difference becomes a source of noise. Amplitude noise is caused by the laser emission process. Noise power increases with rising optical power, reaches a maximum at the threshold level, then decays at higher power levels.

Fiber Noise: Fiber noise specifically relates to modal partition noise. In a multimode laser, competing longitudinal modes cause time-dependent spectral distribution rather than the normal time-average distribution. Because each of these competing modes corresponds to a color, chromatic dispersion of the fiber separates or partitions these modes, causing a reduction in system bandwidth and increased noise. This effect is not present in single-mode fiber, although another phenomenon called modal noise can create similar difficulties when lasers are used with single-mode fibers.

Receiver Noise: The photodiode, the conversion resistor, and the amplifier all contribute to receiver noise. Shot noise is the dominant form of receiver noise. Shot noise, caused by random fluctuations in the current that arise from the discrete nature of electrons, relates to the statistical arrival of the photons at the detector. Receiver noise is directly proportional to the square root of the receiver's bandwidth.

Since the flow of electrons is a random process, a simple counting statistics equation can be used to estimate the noise level. The noise level will be approximately equal to the square root of the number of electrons flowing per unit time. Since a single electron has a charge of 1.6×10^{-19} Coulombs, then 1 Ampere of current is the flow of 6.25×10^{18} electrons per second. The noise associated with that flow is 2.5 x 10⁹ electrons per second or 0.4 nA. The same approach can be used to estimate error in random samplings such as public opinion polls. If 625 people are polled, the margin of error is 25 or 4% of the total.

Time Jitter and Intersymbol Interference: Time jitter is a short-term variation or instability in the duration of a specified interval. Intersymbol interference is the result of other bits interfering with the bit of interest. Where receiver noise is directly proportional to bandwidth, intersymbol interference is inversely proportional to the bandwidth. The eye diagram is a good way to see the effects of time jitter and intersymbol interference.

Figure 10.2: Eye Diagrams

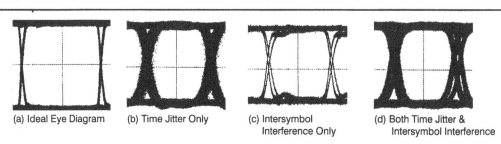

(a) Ideal Eye Diagram   (b) Time Jitter Only   (c) Intersymbol Interference Only   (d) Both Time Jitter & Intersymbol Interference

Figure 10.2a shows an ideal eye diagram that has no jitter or intersymbol interference. The lines are sharp. In Figure 10.2b, noise has been added which causes time jitter. This causes the lines to broaden. Only intersymbol interference has been added in Figure 10.2c, causing the sharp lines; however, multiple lines can be seen. This is due to amplitude distortion of some of the frequency elements. Figure 10.2d shows the combined effects of time jitter and intersymbol interference.

Bit Error Rate: Bit error rate (BER) is the main quality criterion for a digital transmission system. Any of the factors listed in the sensitivity analysis may cause bit errors. The bit error rate of a system can be estimated as follows:

$$\text{Eq. 10.4} \qquad BER = Q\left[\sqrt{\frac{I_{MIN}^2}{4 \cdot N_0 \cdot B}}\right]$$

Where:

N_0 = Noise power spectral density (A²/Hz).
I_{MIN} = Minimum effective signal amplitude (Amps).
B = Bandwidth (Hz).
Q(x) = Cumulative distribution function (Gaussian distribution).

SIGNAL-TO-NOISE RATIO

The signal-to-noise ratio expresses the quality of the signal in a system. It is a ratio of average signal power and total noise and can be written as:

$$\text{Eq. 10.5} \qquad SNR = \frac{S}{N}$$

In this ratio, signal (S) represents the information to be transmitted while noise (N) is the integration of all noise factors over the full system bandwidth. Expressed in decibels, the equation can be written as:

$$\text{Eq. 10.6} \qquad SNR \text{ (dB)} = 10 \cdot \log_{10}\left(\frac{S}{N}\right)$$

BER and SNR are interrelated; a better SNR yields a better BER. Receiver sensitivity is determined by the optical power necessary to achieve a given SNR, and thus a given BER. For instance, a signal-to-noise ratio of 11 dB would yield a BER of about 10^{-9}. Improving the SNR by less than 1 dB to 11.75 dB improves the BER by a factor of 1,000 times to 10^{-12}. The exact relationship between SNR and BER varies depending on the data encoding method. This interdependent relationship, shown in Figure 10.3, is typical for NRZ (non-return to zero) data.

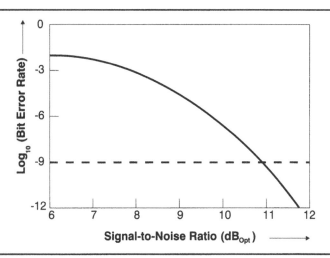

Figure 10.3: BER Dependence on Signal-to-Noise Ratio

MODULATION SCHEMES

The process of passing information over the communication link typically involves three steps: encoding, transmitting, and decoding the information. Another term for this process is modulation scheme. Most communication systems incorporate encoding for a variety of reasons. It may improve the integrity of the transmission or allow more information to be sent per unit time. In some cases, the encoding scheme takes advantage of some strength of the communication medium or overcomes some inherent weakness. Many modulation schemes exist; the four most common modulation schemes are frequency modulation (FM), amplitude modulation (AM), pulse-code modulation (PCM), and digital encoding.

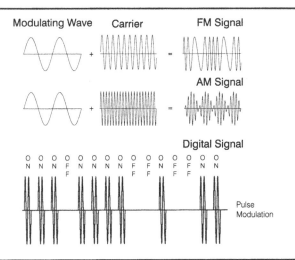

Figure 10.4: Common Modulation Schemes

Amplitude Modulation (AM)

Amplitude modulation or AM is the same scheme used in AM radio. It is a simple technique that often results in very low-cost hardware. There are basically two types of AM techniques: baseband and RF carrier. In a baseband system, the input signal directly modulates the strength of the transmitter output, in this case light. In the RF carrier AM technique, a carrier with a frequency much higher than the information to be encoded is used as the heart of the transmitter. The amplitude of the carrier wave varies according to the amplitude of the information being encoded. In a fiber optic system, the magnitude of the voltage input signal directly translates into a corresponding light intensity. Figure 10.5 shows the function of a typical AM system.

Figure 10.5: Typical AM System

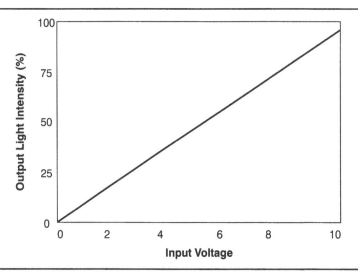

This simple technique has two main drawbacks however. First, the system requires the use of highly linear components throughout so that the signal is not distorted as it travels through the communication link. Second, since the information is encoded by varying light intensity, it becomes difficult to separate real signal level variations from changes in the optical loss in the fiber itself. For instance, providing a 100% maximum signal into the optical transmitter with 10 dB of optical loss between the transmitter and receiver, the receiver would indicate that the signal level is 10%. Without some other means, the receiver cannot distinguish between signal level changes and optical loss changes. One challenge of AM transmission over optical fiber is devising a scheme to compensate for the loss associated with the fiber. This problem may be solved two ways: by taking advantage of some special property of the input waveform (for video, the sync pulse is an invariant that can be used to distinguish optical loss from signal level variation) or using a technique that allows the signal level to be interpreted independently from optical loss. One means of accomplishing this is to send a pilot tone at a high frequency that is above or below the frequency of the information being encoded. The need for highly linear components can erase much of AM's advantage over other techniques because of the expense associated with obtaining highly linear LED's. In spite of the difficulties mentioned above, AM is generally the simplest and least expensive approach to encoding information for transmission over fiber.

Frequency Modulation (FM)

Frequency modulation, called FM, is a more sophisticated modulation scheme. This modulation scheme lends itself to one of fiber's strengths by measuring the signals timing information to properly recover the original signal. FM is also immune to amplitude variations caused by optical loss, one of fiber's weaknesses. The heart of the FM modulator revolves around a high-frequency carrier. Now, instead of changing the amplitude of the carrier, the frequency of the carrier is changed according to differences in the signal amplitude. An advantage of FM systems lies in mathematical analyses that show that the signal-to-noise ratio at the receiver can be improved by increasing the deviation of the carrier. For instance, let us assume that a video signal is to be transmitted, and it has a 5 MHz bandwidth. If we use a 70 MHz carrier frequency and cause it to deviate 5 MHz by applying the video signal then we get about a 5 dB enhancement in the receiver's signal-to-noise ratio compared to an AM system. If we increase the deviation of the carrier frequency to 10 MHz, then the improvement increases to 15.6 dB. Another important advantage of FM is that it eliminates the need for highly linear optical components that are required for AM systems. Often optical systems employing FM encoding refer to the technique as pulse-frequency modulation (PFM). This simply means that the FM signal is limited (converted to digital 0's and 1's) before it is transmitted over the fiber. The result is the same. Generally the modulator is designed so that the frequency of pulses increases as the input voltage increases, but there is nothing magic about this convention.

FM optical systems almost always require more complex electronic circuitry than AM optical systems, but often the total cost is comparable since lower-cost optical components can be used in the FM system. Figure 10.6 demonstrates the operation of a typical FM system.

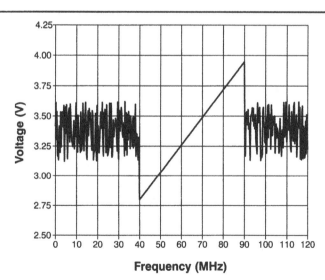

Figure 10.6: Typical FM System Response

Pulse-Code Modulation (PCM)

Pulse-code modulation or PCM converts an analog signal, the human voice for example, into digital format with a number of possible twists. The human voice can be represented by a series of numbers, the value of each number corresponding to the amplitude of the voice at a given instant of speech. Some variants of PCM are listed below. The first three describe analog-to-digital modulation, and the last two are strictly digital modulation schemes.

Pulse-amplitude Modulation (PAM): Information is encoded by a stream of pulses with discrete amplitudes.

Delta Modulation (DM): Pulses are sent at a constant rate with duration determined by the first derivative of the input signal.

Adaptive Delta Modulation (ADM): Similar to DM with the ability to adjust the slope of the tracking signal.

Phase-shift Keying (PSK): Information is sent over a constant carrier frequency. The phase of the carrier is shifted between two levels as determined by the digital bit to be sent.

Differential-phase Shift Keying (DPSK): A variant of PSK that allows for more straight-forward decoding.

One example of this method of digital modulation is outlined in Table 10.5.

Table 10.5: Typical Digital System Operation

Input Signal	Output Code
0 Volts	0000
1 Volt	0001
...	...
9 Volts	1001
10 Volts	1010

ANALOG VERSUS DIGITAL

The world around us is an analog world. Analog means that most variables (e.g., temperature, sound, color and brightness) exhibit a continuous range of difference. Sound is analog in that it continuously varies in both amplitude and frequency within a given range. Another example could be a 60 W light on a dimmer switch. The dimmer adjusts the light within a certain range. The light levels are continuously variable in that there is no one discrete level. The brightness can be adjusted anywhere from completely off to completely on. However, no human sense can distinguish a truly infinite range. There is a point at which two amplitudes appear identical. In hearing, the smallest discernible unit is one decibel.

Digital implies numbers — distinct units. In a digital system all information exists in numerical values of digital pulses. A three-way lamp is an example of a digital system. Each setting brings the bulb up to a specific level of brightness. No levels exist between these three settings. That analog information can be converted to digital and that digital information can be converted to analog is important to electronic communication.

Digital Basics

The bit (short for binary digit) is the basic unit of digital information. This unit has only two values: 1 (one) or 0 (zero). Electronically, the bit is equivalent to the circuit being on or off where 0 = Off and 1 = On. One-bit information is limited to these two values,

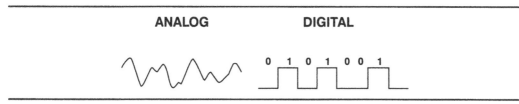

however, if we use two bits of information, more information can be communicated. Look again at a three-way lamp. Using two bits we have 00 = Off, 01 = On (Dim), 10 = Brighter, 11 = Brightest. The more bits used in a unit, the more information can be sent. An 8-bit group is called a byte. A byte gives 256 different meanings to a pattern of 1's and 0's. Each time a bit is added, the number of possible combinations doubles.

Digital systems can be illustrated by pulse trains such as the one seen in Figure 10.8. A pulse train represents the 1's and 0's of digital information. The pulse train can depict high-voltage and low-voltage levels or the presence and absence of a voltage. In dealing with pulses, engineers must consider the shape of a pulse.

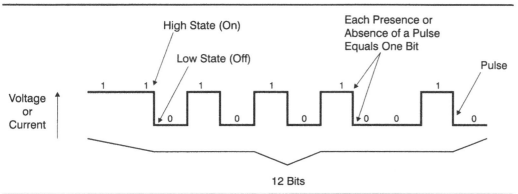

Figure 10.8: Typical Pulse Train

Figure 10.9 illustrates the five main characteristics of a pulse. Amplitude is the height of the pulse. Rise time indicates the amount of time required to turn the pulse on and is typically the length of time required to go from 10% to 90% of amplitude. Rise time is critical because it sets the upper speed limit of the system. The speed at which pulses turn off and on determines the fastest rate at which the pulses can occur. Fall time is the opposite of rise time, representing the time taken to turn the pulse off. Pulse width describes, in units of time, the width of the pulse at 50% of its amplitude. Bit period is the time required for the pulse to go through a complete cycle. Most digital systems are clocked; pulses must occur in the time allotted by the system for a bit period. A clock is an unchanging pulse train that provides timing by defining the bit periods.

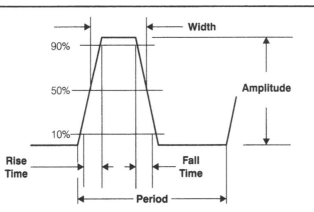

Figure 10.9: Parts of a Pulse

DISPERSION

The mechanisms behind dispersion were discussed in detail in Chapter 3. We will not repeat that detail here; instead, we will focus on the system implications of dispersion. Chromatic dispersion is a basic characteristic of glass fiber that causes wavelengths of light to travel at different speeds through the fiber. Ideally, a transmitter output would be monochromatic and dispersion would not be a concern. However, real lasers used in fiber optic systems output a pulse of light that contains a range of wavelengths. Chromatic dispersion will then cause the pulse to spread out as it travels down the fiber.

As data rates increase, chromatic dispersion rapidly becomes a problem. Chromatic dispersion increases with the square of the data rate and approximately the square of the transmission distance. A system that can transmit a data rate of 10 Gb/s over a distance of 400 km will be limited to a distance of only 25 km (16 times smaller) as the data rate increases to 40 Gb/s. One exception to this is polarization mode dispersion (PMD), which does not increase as rapidly as chromatic dispersion. PMD increases approximately linearly as the data rate increases.

Dispersion Power Penalty

When a digital signal is transported over a long length of fiber, the quality of the eye diagram degrades with distance due to dispersion in the fiber. This degradation coincides with a decrease in receiver sensitivity, expressed in dB as a dispersion penalty.

Increased attenuation, using an optical attenuator with a fiber optic transmitter connected to a fiber optic receiver through a short length of fiber, determines the receiver sensitivity. A given bit error rate (BER), such as 10^{-9} or 10^{-12}, defines the receiver sensitivity limit.

In order to compute the dispersion penalty, we first need to know the spectral width of the laser. For multilongitudinal mode (MLM) lasers, usually Fabry-Perot (FP) type, the spectral width is the root-mean-square spectral width as illustrated in Figure 10.10.

Figure 10.10: MLM Laser Spectral Width

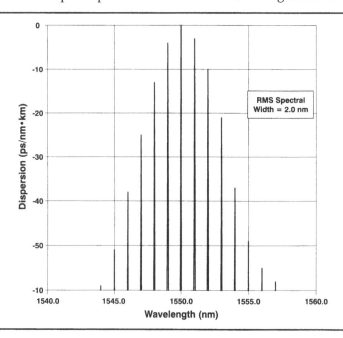

For single longitudinal mode (SLM) lasers, usually distributed feedback (DFB) lasers, the spectral width is the width at the 20 dB down points divided by 6.07. This is the Gaussian spectral width at the 20 dB down point (see Figure 10.11).

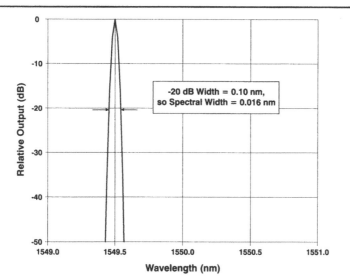

Figure 10.11: SLM Laser Spectral Width

The dispersion power penalty calculation is performed as follows:

$$\tau = \omega \cdot D_\lambda \qquad \text{ps/km}$$

$$f = \log_e(4)/(\tau \cdot \pi) \qquad \text{Hz} \cdot \text{km}$$

Eq. 10.7

$$F_F = f/L \qquad \text{Hz}$$

$$\eta_L = c \cdot (F_R/F_F)^2$$

$$dB_L = 10 \cdot \log_{10}(1 + \eta_L) \qquad \text{dB}$$

Where:

ω = Laser spectral width (nm).
D_λ = Fiber dispersion (ps/nm•km).
τ = System dispersion (ps/km).
f = Bandwidth-distance product of the fiber (Hz•km).
L = Fiber length (km).
F_F = Fiber bandwidth (Hz).
c = A constant equal to 0.5.
F_R = Receiver data rate (b/s).
dB_L = Dispersion penalty (dB).

Figure 10.12 shows the dispersion penalty for a data link operating at the three data rates, 3.11 Gb/s, 1.55 Gb/s and 0.78 Gb/s. The laser is 0.1 nm wide with a center wavelength of 1550 nm and a fiber dispersion of 18 ps/nm•km.

The maximum acceptable dispersion penalty is usually 2 dB. While it is possible for a system to tolerate a larger dispersion penalty if the optical attenuation is low, this is not a recommended course of action. For the example shown in Figure 10.12, the maximum usable fiber length at a data rate of 3.11 Gb/s would be 85 km. At a wavelength of 1550 nm, the optical attenuation would be about 20 dB for that distance, much less than the 30 dB loss budget provided by many high-speed links. In this case, the fiber optic link would be said to be dispersion limited.

Figure 10.12:
Dispersion with a
Normal DFB Laser

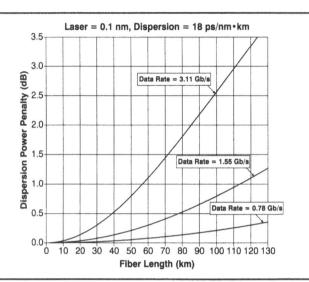

Figure 10.13 shows a second example with a much narrower laser. In this case, all conditions are the same except the laser spectral width is 0.05 nm.

Figure 10.13:
Dispersion with a
Narrow DFB Laser

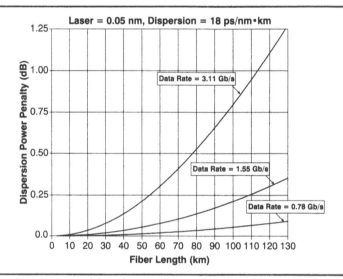

It can be seen that the dispersion penalty has dropped more than a factor of two compared to Figure 10.12. In this case, the dispersion penalty at a data rate of 3.11 Gb/s never reaches 2 dB, even at 130 km. The fiber optic link will, however, reach its optical attenuation limit near this distance. In this case, the fiber optic link is said to be attenuation limited.

Figure 10.14 shows the severe impact that laser spectral width has on the dispersion power penalty of a Fabry-Perot multi-quantum well (MQW) laser with a spectral width of 2 nm. The fiber dispersion remains 18 ps/nm•km at an operating wavelength of 1550 nm.

In Figure 10.14, we have chosen three lower data rates. Even so, the FP laser hits the dispersion penalty limit of 2 dB at a distance of 17 km at 780 Mb/s and 50 km at a data rate of 270 Mb/s. At both of these data rates, the data link is dispersion limited. At a data rate of 100 Mb/s, the link is likely attenuation limited.

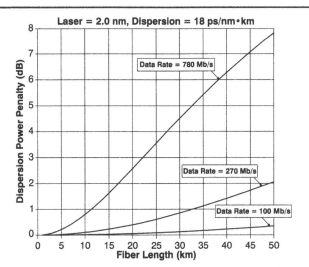

Figure 10.14:
Dispersion with an FP Laser

OPTICAL SIGNAL-TO-NOISE RATIO

In order to transmit a digital signal over a fiber optic transmission system, a minimum optical signal-to-noise ratio (OSNR) must be maintained. Higher data rates require higher OSNR's. For an OC-48 signal, the minimum OSNR is about 15 dB. This increases to about 21 dB at OC-192 data rates. As shown in Figure 10.15, a general topology consists of a number of equal fiber lengths with loss L separated by a total of N EDFA's. When amplifying an optical signal, the original signal gains a finite amount of noise. This noise, called amplified spontaneous emission (ASE), accumulates in optical spans that have multiple optical amplifiers between regenerators. Eventually, as the ratio of the original signal to the added noise lowers, the receiver cannot adequately reconstruct the transmitted signal and errors result. The higher OSNR requirement of higher data rates can impact the overall network design since these signals must consume a greater portion of the available gain at each amplifier.

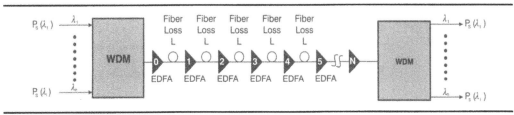

Figure 10.15: OSNR Through a Long Chain of EDFA's and Fiber Lengths

The following expression defines OSNR. Note that this equation applies to spans designed to have equal lengths between in-line amplifiers and equal attenuation.

Eq. 10.8
$$\text{OSNR} = P_S(\lambda_i) - L - N_F - 10 \cdot \log_{10}[h\nu\Delta\nu_0]$$

Where:

$P_S(\lambda_i)$ = Output power per wavelength (dBm).
L = Span loss of the fiber between amplifiers (dB).
N_F = Amplifier noise figure (dB).
N = Number of amplifiers in the chain.
h = Planck's constant.
ν = Optical frequency (Hz).
$\Delta\nu_0$ = Optical bandwidth (nm).

Note the relationships in this expression. The last term, Δv_0, acts as a constant for a given system, and the amplifier noise figure (N_F) acts as a constant for a given optical amplifier. Thus, the critical variables become the number of optical amplifiers in the span, the power per wavelength, and the span loss between amplifiers. From this, it can be seen that every time the number of optical amplifiers in the span doubles, the OSNR degrades by 3 dB. Thus, to compensate, the signal level going into each amplifier must be higher. This is typically accomplished with shorter and shorter spans between amplifiers (reducing the term L). Therefore, more amplifiers allows for shorter spans between amplifiers. Although the system has enough gain to go longer distances, achieving an acceptable OSNR may not be possible. After reaching the OSNR limits of a multiple span, an optical regenerator is required in order to recreate the signal.

Q-Factor

Another measure of signal quality, Q-Factor, is frequently used in modern fiber optic digital transmission systems. It represents the eye diagram's signal-to-noise ratio, and it allows the determination of the minimum bit error rate of an optical network. Determining Q-Factor begins with a standard eye diagram measurement. By taking the average value of the "0" and "1" levels, one can then estimate the standard deviation of the noise associated with "0" level and the "1" level. Q can then be computed using the following equation:

$$\text{Eq. 10.9} \qquad \text{Q-Factor} = \frac{|\mu_1 - \mu_0|}{\sigma_1 + \sigma_0}$$

Figure 10.16 shows the basic parameters that need to be determined in order to compute Q. Using available instruments that automate the measurement of Q, the Q-Factor measurement can be performed very quickly compared to BER measurements.

Figure 10.16:
Determining Q-Factor

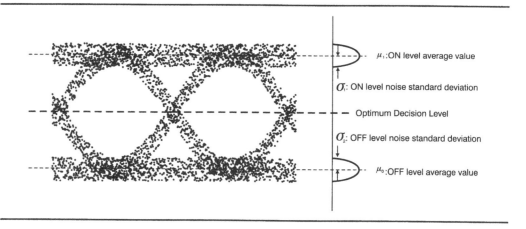

μ_1:ON level average value

σ_1: ON level noise standard deviation

Optimum Decision Level

σ_2: OFF level noise standard deviation

μ_0:OFF level average value

OPTICAL REGENERATORS

A regenerator is an optical receiver that feeds its electrical output directly into an optical transmitter which then launches a new high-power optical signal into the fiber. In many systems, regenerators extend transmission spans by converting incoming photons to electrons, amplifying the electrical signal, retiming and reclocking the amplified signals, and finally converting it back to photons for transmission down the fiber. The number of components needed to do these conversions, as well as jitter reduction, add complexity and expense to this process. Regenerators have the advantage of eliminating the effects of dispersion, nonlinearities and noise, allowing the signal to start out fresh. Figure 10.17 illustrates the typical configuration for using one or more optical regenerators in the optical path.

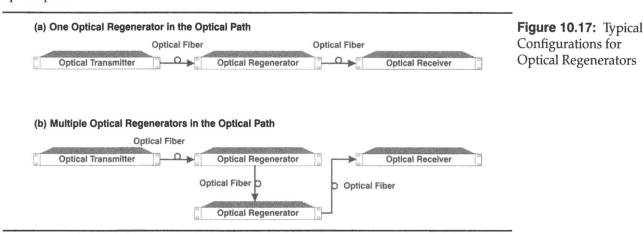

(a) One Optical Regenerator in the Optical Path

(b) Multiple Optical Regenerators in the Optical Path

Figure 10.17: Typical Configurations for Optical Regenerators

COST/PERFORMANCE CONSIDERATIONS

The cost of a fiber optic transmission system can be a critical consideration. While minimizing the cost is often a primary goal, there are performance trade-offs to consider. Component considerations such as light emitter type, emitter wavelength, connector type, fiber type, and detector type will have an impact on both the cost and performance of a system. Common sense goes a long way in designing the most cost-effective system to meet an applications requirements. A properly engineered system is one that meets the required performance limits and margins with little extra. Excess performance capability often means the system costs too much for the specific application.

Understanding the design of a fiber optic system, it is time to turn the discussion to applications for these systems.

Chapter Summary

- The key factors that determine how far one can transmit over fiber are transmitter optical output power, operating wavelength, fiber attenuation, fiber bandwidth, and receiver optical sensitivity.
- The decibel (dB) is a convenient means of comparing two power levels.
- The optical link loss budget analyzes a link to ensure that sufficient power is available to meet the demands of a given application.
- Rise and fall times determine the overall response time and the resulting bandwidth.
- A sensitivity analysis determines the amount of optical power that must be received for a system to perform properly.
- Bit errors may be caused by source intensity noise, fiber noise, receiver noise, time jitter, and intersymbol interference.
- Amplitude modulation (AM), the same scheme used in AM radio, includes baseband and RF carriers and is a simple technique that often results in very low-cost hardware.
- Frequency modulation (FM) is well-suited to the inherent properties of optical fiber since proper recovery of the encoded signals only requires measurement of timing information.
- Analog means continuously changing in value.
- In a digital system all information exists in discrete numerical values.

- The bit is the basic unit of digital information and has only two values: one or zero.
- The five main characteristics of a pulse are rise time, period, fall time, width, and amplitude.
- The dispersion power penalty can contribute to loss of optical signal strength.
- OSNR must be maintained in order to transmit a digital signal over a fiber optic transmission system.
- Q-Factor determines the minimum bit error rate for an optical network.
- Regenerators may be used to reconstruct the optical signal for long distance transmissions.

Selected References and Additional Reading

Alexander, Stephen. "WDM Considerations," from *NFOEC Technical Proceedings, Volume 1.* Proceedings of the National Fiber Optic Engineers Conference, Orlando, FL, Sept. 13-17, 1998.

Baack, Clemens. *Optical Wideband Transmission Systems.* Florida: CRC Press, Inc., 1986.

Chaires, Daryl. "Importance of Mid-stage Access in Next Generation Optical Amplifiers," from *NFOEC Technical Proceedings, Volume 1.* Proceedings of the National Fiber Optic Engineers Conference, Orlando, FL, Sept. 13-17, 1998.

Fuerst, Thomas. "Key Considerations when Implementing OC-192 over DWDM Networks." from *NFOEC Technical Proceedings, Volume 1.* Proceedings of the National Fiber Optic Engineers Conference, Orlando, FL, Sept. 13-17, 1998.

Hentschel, Christian. *Fiber Optics Handbook.* 2nd edition. Germany: Hewlett-Packard Company, 1988

Hersey, Steven and Stan Lumish. "Wanted: More Fiber Capacity." Overland Park, Kansas: Intertec Publishing Corp., 1997.

Palladino, John R. *Fiber Optics: Communicating By Light.* Piscataway, NJ: Bellcore, 1990.

Schweber, Bill. "Optical Amplifiers Literally Pump up the (Photon) Volume." *EDN.* July 16, 1998: 40-44.

Sterling, Donald J. *Amp Technician's Guide to Fiber Optics*, 2nd Edition. New York: Delmar Publishers, 1993.

—. *Universal Transport System Design Guide.* Hickory, NC: Siecor Corporation, 1991.

Vodhanel, Richard S. and J.K. Gamelin. "Large Channel Count WDM Systems: Trade-offs Between Number of Channels, Number of Spans, and Bit Rate," from *NFOEC Technical Proceedings, Volume 1.* Proceedings of the National Fiber Optic Engineers Conference, Orlando, FL, Sept. 13-17, 1998.

Yeh, Chai. *Handbook of Fiber Optics: Theory and Applications.* New York: Academic Press, Inc., 1990.

FIBER OPTIC APPLICATIONS

11

Advantages of fiber optic systems such as light weight, small size, large bandwidth, and EMI immunity make these systems applicable to a wide range of fields and uses.

BROADCAST

The broadcast industry is at the threshold of two major technological revolutions. The first is the rapid move to digitized video. The second is a move to some form of enhanced definition television such as HDTV (high-definition television) or enhanced NTSC (National Television Standards Committee). The combination of these two revolutions make fiber optic technology inevitable. Fiber optic links can support both video and audio broadcast transmissions as well as data transmission. Video transport signal types include multichannel (4, 12, 40, 60, 80 channels are common), point-to-point RS-250, and digitized video (NTSC, CCIR 656, EU95, SMPTE 259). Audio transport signal types include the multichannel audio snake, point-to-point CD quality (stereo), and digitized audio.

Digital Video

As CD technology revolutionized the audio industry, digitized video is revolutionizing the broadcast industry. Because of the bandwidth required to transmit digital video, fiber is the clear choice in this application, especially in the studio environment. However, high levels of compression using standards such as MPEG will undoubtedly be used for mass distribution of digital video. The low bandwidth of MPEG compressed applications, typically 19.4 Mb/s, makes fiber unnecessary.

The first all-digital video broadcast occurred at the 1994 Winter Olympics. Fiber optic links were selected to connect distant outside events such as the downhill ski events and the cross-country ski events to the production studio. The topography of the venues located outside Lillehammer, Norway made microwave links unusable. Serial Digital Video Links, built by Force, Incorporated of Christiansburg, Virginia were used to connect these distant venues to the production studio offering a field test that successfully demonstrated fiber's advantages in this type of application.

Figure 11.1: Serial Digital Component Video Fiber Optic Links

(Photo by Force, Inc.)

The critical fiber parameters for broadcast are light weight, lightning immunity, high bandwidth, long distance, and better signal quality. Actual applications include: intra-studio broadcasting, inter-studio broadcasting, electronic news gathering (ENG), signals to TV camera pan/tilt/zoom pedestals, multimedia distribution systems, and campus video distribution systems.

BROADBAND CATV TRANSPORT

Once dominated by such transmission media as twisted pair, copper coaxial cable, satellite, and microwave transmission, broadband networks are now looking to fiber for the transmission of radio frequency (RF) signals. This transition results from an increased consumer demand for new services, speed, bandwidth, and cost-containment. While all-digital systems may ultimately prevail, they are still prohibitively expensive to install and operate. Recently, hybrid fiber/coaxial cable networks have gained wide acceptance as an alternative to copper-only systems, allowing for a more cost-effective transition. CATV transmission is discussed in further detail in Chapter 12.

Figure 11.2: CATV Video Distribution System

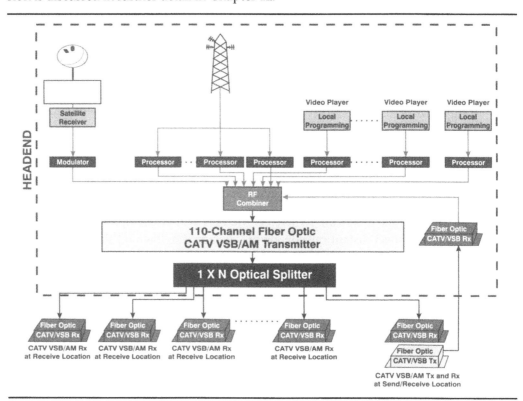

TELECOMMUNICATIONS

The most strenuous demands on fiber optic technology come from the telecommunications industry. A wide range of communications devices employ fiber optic systems. Fiber optics offer greater bandwidth and smaller attenuation. In addition, fiber optic cable costs less than metallic cable and is less susceptible to crosstalk. Fiber was first used in trunk lines that connected central offices with long distance toll centers. But the advantages of fiber to the telecommunications industry, listed below, go beyond supplying phone services.

- Telephone companies can increase system capacity without digging additional underground cable lines.
- Transatlantic fiber optic cable offers continuous communication between North America and Europe.
- Fiber optic cable holds an advantage over satellite communications; satellites are subject to bandwidth limitations, and produce echo and delay effects due to the much longer transmission path. These effects are not created by optical fiber.
- Fiber plays a role in improving the operation of such communications applications as public address (PA) systems, microwave communications, and emergency phone systems.

Fiber-to-the-home and Fiber-to-the-curb

Another telecommunications application gaining wide use is the subscriber loop, a circuit that connects a central office to subscriber telephones. With the high bandwidth of fiber, the telephone company could offer other services such as video and information services. Two approaches are used in subscriber loops. They are fiber-to-the-curb (FTTC) and fiber-to-the-home (FTTH). As each name suggests, in FTTH systems the transceiver is located inside the subscriber's home while transceivers in a FTTC system remain outside the home. Figure 11.3 illustrates the FTTH and FTTC topology.

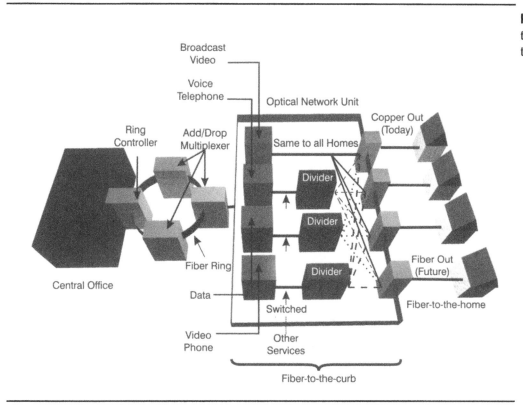

Figure 11.3: Fiber-to-the-curb and Fiber-to-the-home System

As video communications, fiber-to-the-curb and fiber-to-the-home systems increase, the range of applications for this industry will only increase and diversify.

HIGH-RESOLUTION IMAGING

Fiber optics carry high-resolution video images typically encoded as RGB (red-green-blue) signals. Bandwidth can exceed 100 MHz per color. The higher bandwidth and distance capability make this application useful in medical imaging and computer workstations. Other applications include CAD/CAM/CAE computer modeling, air traffic control remote imaging, flight simulation imaging, telemetry, and map making, high-end graphic design, and publishing. Screen resolution and RGB bandwidth are given in Table 11.1.

Table 11.1: RGB Bandwidth and Screen Resolution

Resolution (Non-Interlaced)	Scan Rate (Hz)	Analog RGB Bandwidth (MHz)
640 x 480	60/72	18.4/22.1
800 x 600	60/72	28.8/34.6
1024 x 768	60/72	47.2/56.6
1280 x 1024	60/72	78.6/94.4
1600 x 1280	60/72	122.9/147.5
2560 x 2048	60/72	314.6/377.5

DISTANCE LEARNING (TELE-CLASSROOMS)

Distance learning revolutionizes education, linking classrooms across a campus or a nation. The application involves video/audio links that connect a teacher in one classroom to students in other classrooms. Communication is two-way by video cameras and audio systems connected via fiber optics. The advantages of fiber include no EM radiation, immunity to EMI and EMP, distance capability, and better signal quality. As the cost of a college education skyrockets, the ability to teach more people over greater distance makes education more accessible to those who desire it, and helps the educational facility expand its student base to cover a wider area. Figure 11.4 illustrates the dynamics of distance learning as well as teleconferencing.

Figure 11.4: Distance Learning and Teleconferencing System Configuration

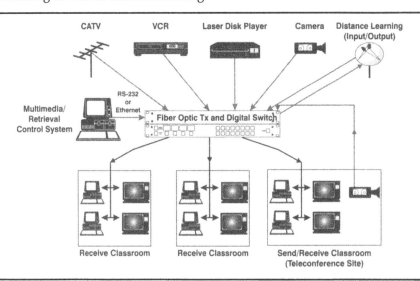

TELECONFERENCING

Similar to distance learning, teleconferencing employs fiber optic systems to connect municipalities and other government units by both video and audio. A specific example of this application, known as electronic magistrate, allows law enforcement officials to arraign suspects via video camera with audio. In rural communities where the arrest site and the courtroom are some distance from each other, this allows officers to process suspects from remotes rather than making a trip to the county seat. The money saved in officer man-hours alone makes this application cost-effective. In addition, teleconferencing connects remote municipalities to the state seat of government and allows business conferences to be conducted face-to-face from a variety of remote locations.

DATA COMMUNICATIONS

Numerous uses for fiber in data communications arise because copper coax cable cannot meet the required bandwidth or data rate. Low-speed data links such as RS-422, RS-485, and RS-232 have found use in security and surveillance industries, distance learning, monitoring systems, and LAN's. High-speed data links capable of data rates up to 3.1 Gb/s and above are finding niches in the computer industry as well as military applications. These important applications deserve closer examination.

Computers

High-speed fiber optic links allow quick transmission of data from mainframe to remote. The bandwidth capabilities of fiber allow for a local area network (LAN) operating at high data rates. Applications in computers fall into two major categories: peripheral interconnection and local area network setup. In both areas, the critical fiber parameters include immunity to EMI and EMP, high bandwidth, and long distance. The low cost of installation and ease of upgrade are factors as well. In terms of computer peripherals, fiber connects one computer to other computers and computers to smart sensors. In local area networks, the use of fiber is more extensive.

Local Area Networks

A local area network or LAN is an electronic communications network that interconnects equipment such as computers, printers, fax servers, modems, and plotters (to name a few) in a limited geographical area like an office building or a campus. All equipment can communicate with other equipment in the LAN. Each point of attachment is a node; each node is an addressable point, capable of sending and receiving information to and from the other nodes. The use of fiber optics is an attractive approach to LAN's for two reasons. It extends transmission distances, and it offers EMI immunity. The two most popular LAN's currently in use are IEEE 802.3 ethernet and IEEE 802.5 token ring. Both systems incorporate fiber optic interconnecting devices but transmit the signals over copper coax twisted pairs. The fiber distributed data interface (FDDI) is the first local area network designed in all aspects to use fiber optics. Its impressive performance offers a data rate of 100 Mb/s over 100 km with up to 1,000 attached workstations. Today, ethernet is specified for speeds up to 1,000 Mb/s.

Network Automatic Protection Switching

Network failures, whether due to human error or faulty technology, can be very expensive for users and network providers alike, making network protection a new challenge for the communications industry. Synchronous networks incorporate a wide range of standardized mechanisms, known as automatic protection switching (APS), in order to compensate for failures in network elements.

Two basic types of protection architecture APS include linear protection and ring protection. Point-to-point connections use linear protection mechanisms. Point-to-multipoint networks use ring protection mechanisms, which can take on many different configurations. Both mechanisms use spare circuits or components to provide the backup path.

Linear protection is also know as 1+1 APS. Here, one protection line exists for each working line. If a defect occurs, the protection agent in the network elements at both ends switches the circuit over to the protection line. A defect, such as LOS (loss of signal) switching at the far end, triggers the switchover to the protection line. This 1+1 APS protection has the advantage that allows implementation at the most basic hardware level. To its disadvantage, 1+1 APS requires two times the hardware required to carry the data.

A more sophisticated form of protection is called 1:N diverse protection structure. In this case a single protection line covers a group of N working lines. Assuming that only one of the working lines will fail at any given time, the 1:N structure design keeps the working lines independent from each other. This 1:N protection requires much less hardware than the 1+1 APS scheme. However, the switching between working systems and the backup system must occur at a higher level in the system. Still, this structure's significant cost advantage overrides other considerations in most cases.

Figure 11.5: SONET Protection Strategy

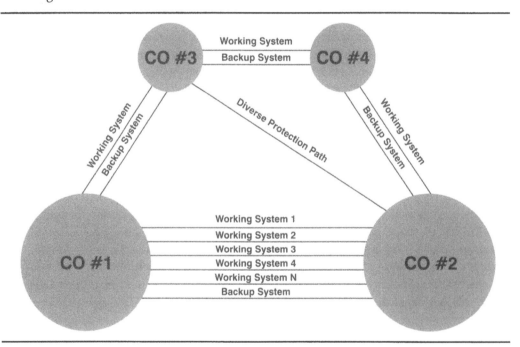

Figure 11.5 shows an example of 1+1 APS protection, 1:N protection, and diverse protection path. The 1:N protection scheme is shown between CO #1 and CO #2. There are N working systems (usually a pair of optical fibers, one for each direction) and a single backup system (usually a single pair of optical fibers, one for each direction). If any of the systems from 1 to N fail, the traffic that would normally be carried by that system is switched to the backup system. Between CO #1 and CO #3, CO #3 and CO #4, and CO #4 and CO #2 are examples of 1+1 APS. In each case, there is a single working system (usually a pair of optical fibers, one for each direction) and a single backup system (usually a single pair of optical fibers, one for each direction). If the working system fails, the traffic that would normally be carried by that system is switched to the backup system. Figure 11.5 illustrates a strategy that includes a backup system between CO #3 and CO #2. The backup fiber runs along a different path than the normal working and protection fibers, offering a hedge against an event that may sever all fibers between two locations.

CONTROL SYSTEMS AND INSTRUMENTATION

Control Systems

Fiber optic technology allows control systems to be integrated through multiplexing signals and connecting modems, and high-speed links. A variety of industries find applications for control systems. For example, in chemical plants, optical fiber senses the physical and chemical parameter changes during a chemical process and activates the control mechanisms. Because the optical sensors are small and impervious to many adverse environments, the sensors can be placed in more than one location and wired in series to provide constant monitoring during a chemical process. In aerospace applications, light weight and EMI/RFI immunity of optical fiber cables replace heavy, EMI sensitive copper cables. Future applications will monitor critical airplane controls during flight. Optical fiber is already used in automobile instrument panel illumination, but the use of sensors to monitor controls is in development. These sensors monitor lights, exhaust emissions, spark plug firing performance, or the firing mixture of the automobile. Radar control systems and traffic control systems that interconnect traffic signals and monitor traffic flows are already in use, and more areas are installing such highway monitoring systems.

Instrumentation

Fiber optics can be utilized in scientific instruments in many ways. Sensing capabilities are important, but size is also a factor, and fiber optic systems can be fitted to almost any system design. Developments such as the endoscope, a variation of the fiberscope, allow engineers to look into the cavity of an engine or into a reactor during operation. Fiber optic interferometric instruments allow a controlled environment and an environment under test to be simultaneously monitored and the differences to be calibrated and measured. Many industries use fiber optic holography; highly accurate three-dimensional images allow these industries to check for flaws in aircraft structure, boiler welding and other objects. Other fiber optic instruments measure velocity.

MILITARY

Fiber optics plays an important role in military applications in a variety of ways. A single optical fiber can replace miles and pounds of copper wire used in many types of control systems. On ships, fiber optic links carry video signals from surveillance cameras to a central control location, eliminating interference caused by powerful radar systems located near these cameras. There are fiber optic radar systems, fiber optic missile launchers, and fiber optic torpedo launchers that allow the missile to be monitored and course corrected for a direct hit. Tactical ground control systems in military LAN's let field posts monitor and gather environmental information and stay in contact with other field stations. Fiber offers a security advantage since optical fiber is much more difficult to tap than metallic wire.

Figure 11.6 shows the data converter set developed to upgrade the Hawk missile system from copper cable to optical fiber. Developed in cooperation with Force, Inc. and Northrop Grumman Corp., the upgrade used a Lucent tactical fiber optical cable assembly (TFOCA) to replace the heavy copper cable used in the original design. The copper cable connected the control unit with the radar system and was limited to 500 feet. This endangered personnel since the Hawk uses an active radar illuminator and is therefore susceptible to radar-homing missiles. The copper-based system also weighed 300 lbs., making it difficult to transport and deploy. Using the data converter set as a transparent drop-in replacement with the tactical fiber optical cable assembly, the maximum distance was increased to 6,500 feet, while reducing the cable weight to only 140 lbs. Force, Inc. and Northrop Grumman Corp. applied mature fiber optic technology to replace the copper links without affecting system functions or redesigning any existing Hawk missile hardware.

Figure 11.6: Fiber Optic Data Converter Set Mounted on Hawk Missile System Hardware

(Photo by Northrop Grumman Corp. and Force, Inc.)

SECURITY AND SURVEILLANCE

Security applications benefit because it is difficult (but not impossible) to tap optical fiber without detection. These systems radiate no electromagnetic energy making non-invasive eavesdropping techniques useless. Optical fiber carries both video and voice surveillance and incorporates data links such as RS-232, RS-485, and access control. Perimeter security, tunnel/highway monitoring, airport security, and access control use fiber optics with excellent results. In the early 1990's airport security requirements became more stringent, requiring electronic card-key access systems to screen the entry of all personnel. These systems implement video monitoring and data access at all gates. An employee's card would access a gate only if the video image being transmitted from that gate to a control center matched the image data stored in a computer. The distances involved, especially in larger airports, make fiber the best choice for these systems.

Figure 11.7: Typical Video/2-Way Data Link for Security and Surveillance Applications

(Photo by Force, Inc.)

Chapter Summary

- The broadcast industry is at the threshold of two major technological revolutions, digitized video and high definition television or enhanced NTSC.

- Because of the bandwidth required to transmit digital video, fiber is the clear choice in this application.

- Because all-digital systems are prohibitively expensive to install and operate, hybrid fiber/coaxial cable networks have gained wide acceptance as an alternative improvement to copper-only systems.

- The telecommunications industry is the heaviest user of fiber optic technology.

- Two approaches of bringing fiber optic technology to the consumer are fiber-to-the-curb (FTTC) and fiber-to-the-home (FTTH).

- High-resolution imaging applications include the endoscope, CAD/CAM/CAE computer modeling, air traffic control remote imaging, flight simulation imaging, telemetry, geophysical computer imaging, map making, and high end graphic design.

- Distance learning links classrooms across a campus or a nation.

- Teleconferencing employs fiber optic systems to connect municipalities and other government units by both video and audio.

- In computers, high-speed fiber optic links allow quick transmission of data from mainframe to remote.

- A local area network (LAN) is an electronic communication network that interconnects equipment such as computers, printers, fax servers, modems, and plotters in a limited geographical area like an office building or a campus.

- Network APS

- All Optical Networks

- Control systems can be integrated using fiber optics by multiplexing signals, modems, and high-speed links.

- Fiber optics can be utilized in scientific instruments such as sensors, interferometric devices, and holographic imagers.

- Military applications of fiber optics technology include fiber optic links on ships that carry video signals from surveillance cameras to a central control location, fiber optic radar systems, fiber optic missile launchers, and fiber optic torpedo launchers.

- Security applications benefit from the use of fiber optics because the links are very difficult to tap without detection.

Selected References and Additional Reading

—. *Basics of Fiber Optics: Applications for CATV.* State College, PA: C-Cor Electronics, Inc., 1991.

Bazaar, Charles. "Fibre Channel: The Future of High-Speed Connectivity." *Fiberoptic Product News.* May 1995: 34-35.

Cottingham, Charles F. "The Fiberoptic Challenge for the '90s and Beyond." *Fiberoptic Product News.* 10th Anniversary Issue 1990: 31-32.

Engineering Staff, Codenoll Technology Corp. *The Fiber Optic LAN Handbook.* New York: Codenoll Technology Corporation, 1990.

Henkemeyer, Rich. "How to Future-Proof a Hybrid Fiber/Coaxial Cable Network." *Lightwave Magazine.* May 1995: 40-41.

Kopakowski, Edward T. "Metropolitan Area Networks for Interactive Distance Learning." *Fiberoptic Product News.* September 1995: 17+.

Kopakowski, Edward T. "Video Cabling Alternatives to Coax in Premise Wiring Environments." *Fiberoptic Product News.* August 1996: 25-30.

Paff, Andy. "Hybrid Fiber/Coaxial Cable Networks to Expand into Interactive Global Platforms." *Lightwave Magazine.* May 1995: 32-38.

Shoemake, Mike and Lynn Woods. "FDDI in the" [sic]. *Fiberoptic Product News.* August 1995: 3-5.

Yeh, Chai. *Handbook of Fiber Optics: Theory and Applications.* New York: Academic Press, Inc., 1990.

VIDEO OVER FIBER

12

Video signals are complicated. A close inspection of a television screen reveals that the picture is comprised of many horizontal lines drawn one after another. The video signal contains the information to draw these lines, detailing whether parts of the line should be dark or light and how to display the colors. Sound is also encoded in the signal.

METHODS OF ENCODING VIDEO SIGNALS

There are three predominant methods of encoding a video signal: amplitude modulation (AM), frequency modulation (FM) and digital modulation. (See Table 12.1.)

Parameter	AM	FM	Digital
Signal-to-Noise Ratio	Low-Moderate	Moderate-High	High
Performance vs. Attenuation	Sensitive	Tolerant	Invariant
Transmitter Cost	Moderate-High	Moderate	High
Receiver Cost	Moderate	Moderate-High	High
Receiver Gain Adjustment	Often Req.	Not Req.	Not Req.
Installation	Adjustments Req.	No Adjustments Req.	No Adjustments Req.
Multichannel Capability	Good Capability Req. High Linearity Optics	Fewer Channels	Good
Performance Over Time	Moderate	Excellent	Excellent
Environmental Factors	Moderate	Excellent	Excellent

Table 12.1: Comparison of AM, FM and Digital Encoding Techniques

The difference between various modulation schemes can be understood by examining their corresponding frequency spectra. The very simple baseband AM occupies the region from DC to about 5 MHz and requires the least bandwidth (assuming we are talking about uncompressed digital encoding techniques). The RF carrier modulation spectrum is similar; it has been shifted to a non-zero frequency (F). This approach requires additional bandwidth and offers no advantage over baseband operation in a system where a single channel is carried on each fiber. However, it allows multiple channels to be combined onto a single fiber. With vestigial-sideband AM, the spectrum again shifts to a non-zero frequency (F), and filtering removes the lower sideband. It allows for more efficient use of the spectrum as compared to straight RF carrier AM, requiring half the bandwidth per channel. The presence of harmonics yields the notable difference between sine wave FM, square wave FM, and pulse-frequency modulation. The square wave FM spectrum signal contains only odd-order harmonics. The pulse-frequency modulation spectrum contains all odd- and even-order harmonics yielding a cluttered spectrum poorly suited for multiple-channel stacking; however, it retains its value as a single-channel transmission scheme.

Like sine wave FM, RF carrier AM and vestigial-sideband AM lack harmonics, making them suitable for multiple channel transmission. Multiple channels can be combined for transmission over a single fiber by assigning different carrier frequencies to each video signal. The resulting modulated carriers are summed to yield a single composite electrical signal. For instance, a four channel sine wave FM system could occupy the frequencies of 70 MHz, 90 MHz, 110 MHz and 130 MHz.

VIDEO QUALITY PARAMETERS

There are dozens of parameters used to describe the quality of a video signal. The ones encountered most often are signal-to-noise ratio (SNR), differential gain (DG), and differential phase (DP). The signal-to-noise ratio measures the clarity and crispness of the picture. A sharp clear picture has high SNR. Conversely, a picture that shows lots of snow or other interference has a low SNR. Differential gain measures the portion of the video signal that controls how bright a given dot will be. A perfect link with zero DG distortion shows everything in the picture at exactly the right brightness. However, a link with a large amount of DG distortion shows incorrect shades of brightness. Differential phase measures the portion of the video signal that controls the color, hue, or shade. A link that has zero DP distortion shows all colors exactly right. while large amounts of DP distortion cause incorrect colors.

ADVANTAGES OF FM

In an FM system, link loss remains independent of the received signal amplitude, whereas in an AM system the received signal amplitude directly reflects the link loss. This dictates that the AM link must have gain control at the receiver. Sometimes, this is a manual control because of the difficulty of implementing an automatic gain control (AGC) on recovered baseband video. This complicates the installation of some AM links because the user must inject a test signal and use an oscilloscope to set the correct receiver gain level. It also means that the link will not automatically adapt to changes in the link loss. Link loss could change over time due to fiber degradation or transmitter power variations. The received optical power will almost certainly change when optical connectors in the link are demated and remated. Ideally, a practical AM system requires an AGC.

A picture containing a range of levels from black to white will be 140 IRE units high, which also corresponds to 1.0 Vp-p. Thus, one IRE unit equals about 7.14 mV. A picture containing only black levels will only be 40 IRE units high, or about 0.286 Vp-p. Using a simple AGC to output 1.0 Vp-p yields an severely distorted output waveform for an all-black picture. The sync-pulse, usually 40 IRE units high, will become 140 IRE high, and most monitors cannot interpret the signal. Therefore, a simple peak-to-peak AGC cannot be effective in setting the gain of an AM receiver. A more sophisticated circuit that analyzes the input waveform and sets the sync pulse level to 40 IRE units is required.

The drawback to this approach is that the circuit has to know what picture format to expect. It has to assume that the picture format is NTSC, set by the National Television Standards Committee, PAL, an acronym for phase line alteration, or another type. A circuit designed to work with NTSC may not be compatible with other formats. Long-term expandability of the link is limited, for example, NTSC-compatible-only AM links today will be obsolete when the change over to high-definition television (HDTV) occurs.

FM transmission can be independent of the encoding format. The FM modulation/demodulation technique provides proper recovered signal level independent of link loss and the input video format. The recovered signal level varies only by component tolerances in the transmitter and receiver. The output level typically equals the input level plus or minus a small percentage. This reduces long-term maintenance and ensures proper performance and consistent quality over the duration of the link's life.

FM can yield a higher signal-to-noise ratio (SNR) than AM under similar link loss conditions. In a comparison of the estimated SNR for an AM versus FM scenario, each modulation scheme can be summarized by a single factor relating the video SNR to the fiber optic link carrier-to-noise. The FM β is unity for this comparison; a larger value of β would result in improved SNR performance. Additional improvements would be obtained for β values ranging from one to ten.

Of course the penalty paid for larger β values is a large increase in transmission bandwidth. With all other factors equal, an FM system with a β of two has about an 8 dB advantage over AM. This translates into either longer distance links, more loss margin, higher picture quality, or some combination of the three factors. This analysis overlooks nonlinearities in the link and their effect on picture quality. Nonlinearities in AM fiber optic links can cause cross-modulation products that result in lines and bars in the transmitted picture. While this interference is not noise, it can be quite annoying.

DISADVANTAGES OF FM

FM's key disadvantage is the requirement of more bandwidth. If one considers an NTSC video signal with a 5 MHz bandwidth allotment, then a baseband AM system only requires 5 MHz of bandwidth. However, an FM system requires at least seven times this bandwidth for a total of 70 MHz to accommodate the 70 MHz FM carrier. This requires higher bandwidth from the electro-optical components and the associated electrical components.

A good quality 62.5/125 μm fiber has a bandwidth-distance product of at least 400 MHz•km. Thus the carrier frequency of 70 MHz becomes a factor at a distance of 5 km or longer. It becomes a problem at 1.5 km distance when 850 nm LED's are used. Multimode fiber used with an 850 nm LED has very limited bandwidth because of high dispersion.

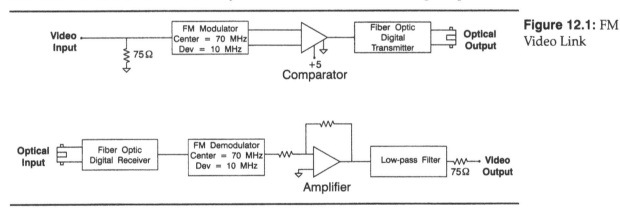

Figure 12.1: FM Video Link

Figure 12.1 shows the typical architecture of an FM video link. The top of the figure illustrates an FM transmitter. In this design, the FM modulator centers around 70 MHz with a deviation of 10 MHz. Once through the modulator, the comparator converts the analog signal to a digital signal, and the fiber optic digital transmitter converts the signal into light pulses. The FM video receiver operates in the reverse manner, converting the digital signal to an electronic signal. The FM demodulator decodes the signal and then removes the carrier frequency components using a low pass filter.

FM VIDEO LINK PERFORMANCE

Figure 12.2 illustrates the performance typically achieved by a fiber optic video link showing signal-to-noise ratio of the video signal versus fiber length in meters. The signal quality stays level over up to 9,000 meters of fiber and rolls off to 16,000 meters where it drops below the minimum specification of 48 dB signal-to-noise ratio.

Figure 12.2: SNR versus Fiber Length

(Data based on the performance of Force, Inc. fiber optic FM video links.)

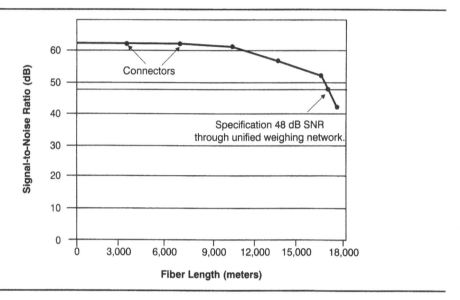

Figure 12.3 demonstrates FM video links as compared to AM video links. At 0 dB optical loss, AM has a higher 70 dB signal-to-noise ratio, but at 3 dB of optical loss, the AM signal-to-noise ratio drops below the FM SNR. This continues for the remainder of the optical loss range. At a practical (and typical) loss of 10 dB, the FM link has a 13 dB signal-to-noise advantage over the AM link. The slope of the AM video link indicates that the signal-to-noise ratio drops 2 dB every time the optical loss increases by 1 dB. This contrasts to the FM link that shows nearly flat signal-to-noise ratio performance over the usable loss range. At the loss limit, the signal-to-noise ratio drops by 2 dB per 1 dB of optical loss like the AM link.

Figure 12.3: FM Video Link versus AM Video Link

(FM data based on the performance of Force, Inc. fiber optic FM video links.)

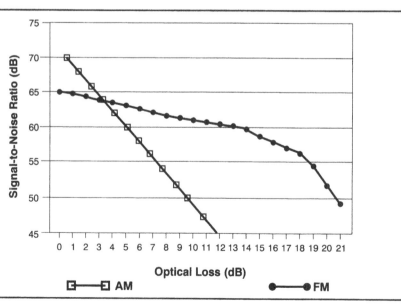

DIGITAL VIDEO

The conversion of analog video inputs to digital values allows the digital transmission of video signals over optical fiber. At the transmitting end, analog-to-digital (A/D) converters encode the baseband video channels. Next, the digital channels are time-division multiplexed (TDM) and sent to the laser transmitter. The digital signal is converted into light pulses; the laser is on for a one and off for a zero. At the receiving end, the light pulses are converted back into electrical pulses. The pulses are time-division demultiplexed (TDD) and sent through a digital-to-analog (D/A) converter. This converts the information back into a baseband video signal. Figures 12.4 and 12.5 illustrate a digital optical transmission system for sending and receiving 16 digitized video and audio signals.

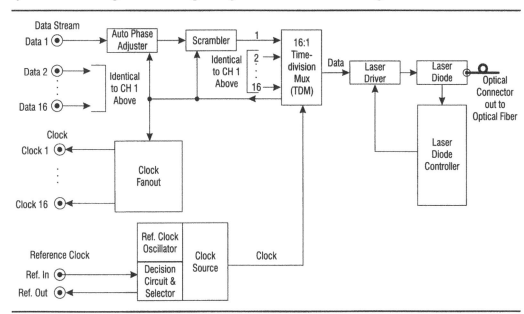

Figure 12.4: Sixteen Channel Configuration for an Optical Transmitter Multiplexing & Sending Sixteen Digitized Video Signals

(Comlux® transmitter block diagram courtesy of Force, Incorporated.)

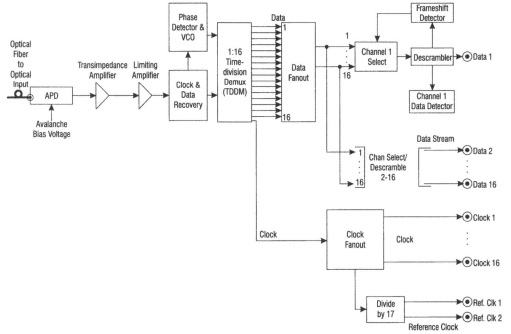

Figure 12.5: Sixteen Channel Configuration for an Optical Receiver Receiving & Demultiplexing Sixteen Digitized Video Signals

(Comlux® receiver block diagram courtesy of Force, Incorporated.)

The analog to digital conversion requires three steps: sampling the signal at regular time intervals, quantizing the signal into discrete bins, and coding the quantized values into an appropriate binary code.

Digital video has several inherent advantages where video transmission is concerned. Digital systems do not require linear light sources, allowing the system to have a wide range of non-critical operating parameters. They can be used with both single-mode or multimode fiber, have a high immunity to noise, and can be retransmitted numerous times without signal degradation.

Several standards exist or have been proposed for digitized video transmission. All result in a serial digital data stream. The most important standards are shown in Table 12.2. Of the standards listed, the first five are currently the most active. EU95, a digitized HDTV standard, is the focus of several European Community development programs. The SMPTE (Society of Motion Picture and Television Engineers) Committee recently released a second digitized HDTV standard, SMPTE 292M.

One cannot help but marvel at how much bandwidth these standards require compared to standard broadcast analog video's bandwidth of 4.5 MHz. Digital compression will help with the distribution of these signals. Studios will use full bandwidth or minimally compressed (lossless) signals that will require the use of fiber as the prevalent means of signal distribution. Copper coax is worthless at gigabit data rates.

Table 12.2: Digital Video Formats

Standard	Picture Aspect Ratio	Sample Rate	Word Rate	Line Serial Date Rate
SMPTE 259M Level "A" (NTSC)	4 x 3	14.318 MHz	14.318 M/s	143.18 Mb/s
SMPTE 259M Level "B" (PAL 4 f_{sc})	4 x 3	17.7 MHz	17.7 M/s	177 Mb/s
SMPTE 259M Level "C" (4:2:2)	4 x 3	13.5 MHz	27 M/s	270 Mb/s
SMPTE 259M Level "D" (4:2:2)	16 x 9	18.0 MHz	36 M/s	360 Mb/s
ITU-R601, CCIR 656	4 x 3	13.5 MHz	27 M/s	270 Mb/s
EU95	16 x 9	72.0 MHz	144 M/s	1440 Mb/s
SMPTE 292M	16 x 9	74.25 MHz	148.5 M/s	1485 Mb/s

SERIAL DATA TRANSMISSION FORMATS AND STANDARDS

All of the standards in Table 12.2 describe the conversion process from an analog video signal to a digitized signal. The standards describe the scale factors, word formats, overall data structure, and how to serialize the data for copper coaxial cable transport. The first definition is in SMPTE 259M, "10-Bit 4:2:2 Component and 4fsc Composite Digital Signals: Serial Digital Interface." SMPTE 297M, "Serial Digital Fiber Transmission for ANSI/SMPTE 259M Signals" was released later to standardize the fiber optic interface for transmitting SMPTE 259M signals.

Currently, broadcasters are in the process of changing over to digital broadcasting and HDTV resolutions which require several new SMPTE standards. SMPTE 292M is similar to SMPTE 259M, but the data rates are much faster, and the standard addresses fiber optic specifications. The 1.485 Gb/s data rate (full bandwidth HDTV) allows for copper coaxial cable lengths of 100 meters while fiber is capable of greater than 100 kilometers. SMPTE 310M for HDTV formats using MPEG-2 data compression at rates up to 40 Mb/s are also being implemented over fiber optics. Both SMPTE 292M and 310M signals are being used by digital broadcasters in their studio-to-transmitter (STL) links over fiber. Table 12.3 lists the key parameters of the serial digital data link associated with these SMPTE standards.

Parameter	259M Requirement	292M Requirement	310M Requirement
Input/Output Load	75 Ohm	75 Ohm	75 Ohm
Return Loss (Electrical)	>15 dB from 5 MHz to F_{CLOCK} (Max. 360 MHz)	>15 dB from 5 MHz to F_{CLOCK} (Max. 1.485 GHz)	>30 dB from 100 kHz to F_{CLOCK} (Max. 70 MHz)
Data Amplitude (Into 75 Ohms)	800 mV ±10% p-p	800 mV ±10% p-p	800 mV ±10% p-p
DC Offset	0.0 ± 0.5 Volts	0.0 ± 0.5 Volts	0.0 ± 0.5 Volts
Rise/Fall Times (20-80%)	> 0.4 < 1.50 ns	≤ 270 ps	> 0.4 < 5.0 ns
Jitter & Distortion	<±250 ps	≤ 135 ps	2 ns p-p Max.
Electrical Connector	BNC	BNC	BNC

Table 12.3: Signal Levels and Specifications for SMPTE 259M, 292M and 310M

CATV TRANSMISSION

Cable television has its roots in community antenna television and retains the name CATV, the acronym of its predecessor. As consumer demands for services increase, bandwidth and signal quality requirements for cable television transmission also increase, and optical fiber plays a major role in this area of video transmission. While cable systems began as strictly copper systems, fiber backbones for CATV networks are commonplace. The hybrid fiber/coaxial cable (HFC) network has become widespread in the CATV industry. This network uses a fiber optic trunk line as the backbone and coaxial cable lines for the drops to individual sites.

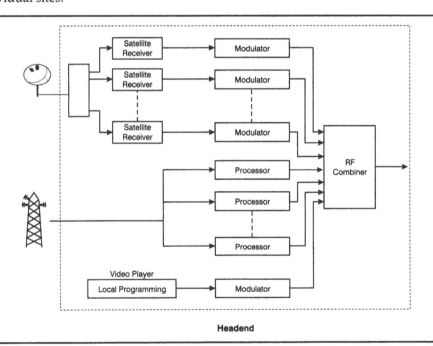

Headend

Figure 12.6: CATV Headend Block Diagram

A CATV system consists of several components. The first component is the cable. As mentioned, copper cable's prevalence declines as bandwidth demands make optical fiber the only choice. Passive components such as signal dividers, signal combiners, and taps create multiple paths in the network allowing greater coverage for more users. Amplifier stations along these paths recover the signal strength after it has been attenuated by the cable. These stations are placed in the transmission line at approximate intervals.

There are three layers of a CATV system. The trunks cascade one after another to cover greater distances. The feeder reaches directly from the trunk to uniformly cover smaller areas, and the drop is the final short-distance connection to the individual receiver, i.e., the cable customers's television. Signals are collected from many sources including satellites and distributed from the headend of the system. The signals are amplified and converted to the desired television channel frequency before being distributed to the cable system.

A few years ago, many predicted that AM CATV over fiber was a dying field. Just the opposite has happened. AM CATV fiber optic links are thriving today. They are perhaps more popular than ever. For many years, 1310 nm was the workhorse for CATV. Much refinement went into the lasers to improve their linearity, lower their noise and increase their output power. Today, optical output powers are exceeding 20 mW (+13 dBm), and linearity is good enough for 110 channels. More recently, 1550 nm has become very popular. The advantage here is that the transmission distance can be increased from 20-30 km to 60-100 km.

Chapter Summary

- Three common parameters used to describe video quality are signal-to-noise ratio (SNR), differential gain (DG), and differential phase (DP).
- The three predominant methods of encoding a video signal are amplitude modulation (AM), frequency modulation (FM), and digital modulation.
- RF carrier AM, vestigial-sideband AM, and sine wave FM are best suited for multichannel transmission of analog signals.
- The inherent advantage offered by FM over AM is that the output signal amplitude is independent of link loss, whereas the uncompensated AM received signal amplitude directly reflects the link loss.
- FM's key disadvantage is the requirement for more bandwidth.
- The conversion of analog inputs to digital values allows for digital video transmission over optical fiber.
- Some inherent advantages of digital video systems are that they do not require linear light sources, can be used with both single-mode or multimode fiber, and have a high immunity to noise.
- Digital video formats include NTSC, PAL (4 f_{sc}), SMPTE 259M (4:2:2), CCIR 656, SMPTE 292M, and EU95.
- A CATV system includes signal dividers, signal combiners, taps, and amplifier stations.
- The three layers of a CATV system are the trunk, the feeder, and the drop.

Selected References and Additional Reading

—. *Basics of Fiber Optics: Applications for CATV.* State College, PA: C-Cor Electronics, Inc., 1991.

Engineering Staff, Codenoll Technology Corp. *The Fiber Optic LAN Handbook.* New York: Codenoll Technology Corporation, 1990.

Fibush, David K. "Video Compression Overview." *Video Systems.* December 1995: 32+.

Henkemeyer, Rich. "How to Future-Proof a Hybrid Fiber/Coaxial Cable Network." *Lightwave Magazine.* May 1995: 40-41.

Kopakowski, Edward T. "Metropolitan Area Networks for Interactive Distance Learning." *Fiberoptic Product News.* September 1995: 17+.

Paff, Andy. "Hybrid Fiber/Coaxial Cable Networks to Expand into Interactive Global Platforms." *Lightwave Magazine.* May 1995: 32-38.

DATA OVER FIBER

13

After discussing the intricacies of video transmission over fiber, we will now look at data transmission where fiber excels because of high bandwidth, low loss, and EMI/RFI/lightning immunity. These characteristics are more important in some applications than in others. In security applications, EMI/RFI/lightning immunity is the paramount consideration. For very long distance applications, low loss prompts the choice for fiber. For high-speed applications, such as gigabit data rates between computers, the high bandwidth determines the use of fiber. After the initially deciding to use fiber optics for a data transmission requirement, other decisions must follow such as:

- Operating wavelength
- Optical loss budget
- Fiber type
- Fiber size
- Optical connector type

DIGITAL DATA TRANSMISSION

Whereas analog transmission over fiber often uses a carrier transmission, digital data transmission typically uses baseband transmission. This means that a logic 0 is sent as a low light level (or the light may be completely off) and a logic 1 is sent as a high light level. Often, employing some sort of data coding (e.g., 4B5B, 8B10B, etc.) guarantees minimum transition density. This coding scheme alleviates the difficulty in designing a fiber optic data link that includes true DC or steady-state data rates requires this data coding. Low data rates (typically <1 MHz) accommodate true DC, but achieving this becomes increasingly difficult at higher data rates, mainly because of the relatively poor DC performance of amplifiers capable of high frequencies.

True DC response is difficult because negative light does not exist. Light is an unbalanced transmission media. When transmitting data using an electrical signal on copper cable, a balanced signal provides optimum signal fidelity, especially at higher data rates. A logic 0 might be sent using a -1.0 Volt level, while a logic 1 might be sent using a +1.0 Volt level. This is a balanced, symmetrical transmission scheme. Recovering the data requires comparing the received signal to a threshold of 0.0 Volts regardless of how much the signal is attenuated during transmission. For instance, if the signal was attenuated by a factor of ten, the received voltage levels would be +0.1 Volts and -0.1 Volts. A threshold of 0.0 Volts would still be perfectly centered allowing optimum data recovery.

DC data transmission with light is not quite so easy. Since negative light cannot be generated, the best one can do is use a zero light level for a logic 0 and a high light level for a logic 1. This makes recovering the data very challenging at the receiver. The attenuation of light through the fiber requires an adjustment of the receiver threshold to 50% of the peak received light. However, the receiver cannot determine the peak if a steady logic 0 is sent. To correct this problem, set the logic threshold to 5% of the largest peak signal. This leads to varying duty cycle distortion as the input light level changes. DC coupled data links also cannot tolerate as much optical loss as an AC coupled data link.

AC coupling the receiver with a capacitor prevents removes the average DC voltage and sets the threshold to a constant 0.0 Volts. The value of this capacitor sets the low-frequency data rate of the link. Typically, the low-frequency limit is set at 100 to 1,000 times lower than the lowest data rate. For instance, the low frequency limit of a 10-200 Mb/s data link is 1 kHz to 10 kHz. Generally, the transmitter is DC coupled. Tailoring the upper frequency limit of the transmitter and receiver to match the data rate range also ensures proper operation of the link. If the upper frequency limit is too low, the recovered data will exhibit intersymbol interference. If the upper frequency limit is too high, excessive noise will degrade sensitivity.

DATA CODING

Data coding takes an unknown, unpredictable data stream and converts it into a data stream with a predictable transition density. An AC coupled receiver must have constant transitions at its input for proper operation. There are two types of coding schemes used, cyclic bit scrambling and block coding. Consider a fictitious block coding scheme 3B6B. Table 13.1 shows how this scheme might work. The input data stream is divided into 3-bit groups, forming the input code. The lookup table then converts the 3-bit input code into a 6-bit output code. To its disadvantage, this scheme generates twice as many bits, doubling the data rate, but it also guarantees a constant transition density of 50%. Even with an input data stream of nothing but zeroes, the scrambled output code has an average duty cycle of 50%. Real scrambling codes are less efficient. Block coding schemes such as 4B5B or 8B10B increase the bit rate by 25%. The alternative data scrambling technique, cyclic bit scrambling, does not increase the bit rate, but it also does not provide as much confidence in the output duty cycle. Chapter 14 discusses leading edge data formats.

Table 13.1: 3B6B Coding Scheme

Input Code	Output Code
000	100011
001	100101
010	100110
011	101001
100	101010
101	101100
110	110001
111	110010

COMMON DATA INTERFACE CHARACTERISTICS

Data links use a variety of interfaces, including TTL, CMOS, RS-485, ECL, and PECL. Table 13.2 lists these along with their associated key attributes.

Table 13.2: Common Data Interface Characteristics

Interface Type	Logic Low Voltage	Logic High Voltage	Receiver Threshold	Transmission Impedance	Frequency Range
CMOS	0.0 Volts	+5.0 Volts	+2.5 Volts	N/A	0-200 MHz
TTL	0.4 Volts	+3.4 Volts	+1.4 Volts	N/A	0-200 MHz
RS-485	-0.2 Volts	+0.2 Volts	+0.0 Volts	120 Ω	0-2.5 MHz
ECL	-1.8 Volts	-0.8 Volts	-1.3 Volts	50 Ω	0-2,000 MHz
PECL	+3.2 Volts	+4.2 Volts	+3.7 Volts	50 Ω	0-2,000 MHz

Table 13.2 shows typical values. Refer to manufacturers' data sheets and published specifications for full details.

KEY DATA LINK ATTRIBUTES

The key parameters associated with fiber optic data transmission are:

- Data rate
- Bit error rate (BER)
- Eye diagram parameters
- Jitter
- Protocol

The fiber optic data link manufacturer designs the data rate limitation. Signal-to-noise limitations determine the BER. Because noise acts as a Gaussian function, the greater the signal-to-noise ratio, the lower the bit error rate. However, when the signal-to-noise ratio drops, errors will increase. Table 13.3 shows the relationship between signal-to-noise ratio and BER. (Note that this relationship depends on the exact coding scheme being used and several modulation parameters.) The relationship becomes very steep at optical signal-to-noise ratios above 10 dB. At an optical SNR of 11.0 dB, BER equals 8.87×10^{-11}. Increasing the optical SNR to 12.0 dB improves the BER to 8.64×10^{-14}, an advancement by a factor of 1,000 times.

The required BER depends on the application and, to some extent, the data rate. For instance, specifying a BER of 10^{-12} for a 2400 Baud RS-232 link would mean that one error occurs every 13.2 years, making an acceptance test impossible to perform. However, if the data rate increases to 2.5 Gb/s, then one error would occur every 6.7 minutes. Table 13.3 contains information that helps to predict very high BER values. For instance, to observe a BER 10^{-3}, increase the optical loss to the appropriate number of dB per Table 13.3 until the desired BER is obtained. Typically, low-speed data links (< 10 Mb/s) use a BER of 10^{-9}. Gigabit data links require a BER of 10^{-12}, and a few requirements have pushed the BER requirement to 10^{-15}.

SNR_{OPT}	BER
6.0	2.13×10^{-2}
6.5	9.80×10^{-3}
7.0	3.49×10^{-3}
7.5	9.61×10^{-4}
8.0	2.05×10^{-4}
8.5	3.38×10^{-5}
9.0	4.32×10^{-6}
9.5	4.27×10^{-7}
10.0	3.27×10^{-8}
10.5	1.94×10^{-9}
11.0	8.87×10^{-11}
11.5	3.15×10^{-12}
12.0	8.64×10^{-14}
12.5	1.84×10^{-15}
13.0	3.02×10^{-17}

Table 13.3:
Relationship Between $SNR_{Optical}$ and BER

The eye diagram offers a powerful tool for analyzing the overall quality of a fiber optic data link. Eye width and eye height are critical parameters, as illustrated in Figure 13.1.

Figure 13.1: Typical Eye Diagram

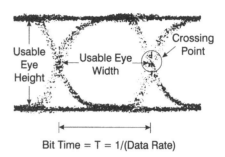

Bit Time = T = 1/(Data Rate)

To observe an eye diagram, an oscilloscope monitors the receiver data output and triggering on the data clock. Usually the data clock is available from the bit error pattern generator. A pseudorandom data pattern generates the worst case eye diagram. Noise (jitter) and intersymbol interference produce the blurry outline seen in Figure 13.1. Noise in the receiver is converted to jitter. Noise and jitter increase as the optical input to the receiver decreases. Non-flat gain somewhere in the fiber optic link generates intersymbol interference. A pseudorandom sequence will contain a wide range of frequency components up to the actual data rate. If the link has flat gain across this entire range, no intersymbol interference will be seen.

The key features of the eye diagram are the crossing points and the open area in the eye. Ideally, the crossing points should be symmetrical and centered, and the open area should be as large as possible. The distance between the crossing points is the reciprocal of the data rate. The other key eye diagram attributes are the usable eye opening width (time) and the usable eye opening height (voltage). Since noise is Gaussian in nature, overtime, the eye will gradually close. The exact values of the eye width and height depend on the desired BER contour. Figure 13.2 shows the BER contours associated with the eye diagram shown in Figure 13.1. Contour A corresponds to a BER of 10^{-6}, contour B corresponds to a BER of 10^{-9}, and contour C corresponds to a BER of 10^{-12}. Usable eye width and usable eye height decrease as the BER requirement gets smaller and tougher to meet.

Figure 13.2: BER Contours of a Typical Eye Diagram

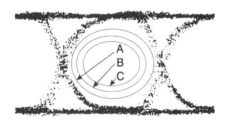

Table 13.4 shows the conversion factor to convert root mean square (RMS) jitter to peak-to-peak jitter. RMS jitter is measured and then converted to peak-to-peak jitter for a given BER. This allows the usable eye to be extrapolated to very small BER values. Jitter increases as the square root of sum of squares while data dependent jitter (intersymbol interference) increases linearly.

BER	SCALE FACTOR
10^{-4}	7.438
10^{-5}	8.530
10^{-6}	9.506
10^{-7}	10.398
10^{-8}	11.224
10^{-9}	11.994
10^{-10}	12.722
10^{-11}	13.411
10^{-12}	14.069

Table 13.4: Conversion for Root Mean Square Jitter to Peak-to- Peak Jitter

Figure 13.3 shows an example of how jitter adds across three circuit elements.

Figure 13.3: Jitter Buildup

The total jitter for the system shown in Figure 13.3 is as follows:

Assume the BER is 10^{-9}. The total system random jitter is:

Eq. 13.1
$$\sqrt{10^2 + 15^2 + 5^2} = 18.7 \text{ ps RMS}$$

The total system data dependent jitter is:

Eq. 13.2
$$25 + 35 + 15 = 75 \text{ ps}$$

Now if the data rate is 500 Mb/s, then the bit time is 2,000 ps. Use the information in Table 13.4 to convert RMS jitter to peak-to-peak jitter:

Eq. 13.3
$$\text{Jitter}_{p-p} = 11.994 \cdot 18.7 = 224 \text{ ps}$$

So now the total jitter is:

Eq. 13.4
$$224 + 75 = 299 \text{ ps}$$

The usable eye width will be:

Eq. 13.5
$$2000 - 299 = 1701 \text{ ps}$$

DISTORTION IN FIBER OPTIC DATA LINKS

During transmission, the ideal data link, fiber optic or otherwise, introduces no data distortion, but why does distortion matter in a digital data link? Isn't it just a matter of ones and zeroes? Yes and no. If the data link introduces excessive distortion, the ability to recover the ones and zeroes will be hampered. Figure 13.4 shows how noise hampers the ability to recover the data. The noise closes the eye diagram which makes data recovery more difficult. The key types of distortion are noise (jitter), data dependent distortion, and duty cycle or pulse width distortion. Noise can be measured in a digital link as random jitter (RJ) that occurs in the receiver, discussed in Chapter 6. Random jitter is simply the time-domain manifestation of voltage noise. The topology of a fiber optic receiver was described in Chapter 6. Figure 13.4 shows a typical receiver topology and the relationship between voltage noise at the transimpedance amplifier output and the random jitter at the output of the decision circuit. As the amplitude of the voltage noise increases, the amount of time-domain random jitter will increase as well.

Because random jitter, like noise, behaves as a Gaussian function, it is measured as a root mean square (RMS) value. The information in Table 13.4 can be used to convert the RMS values to peak-to-peak (p-p) values. Note that the exact scaling factor depends on the BER. Random jitter can be further broken down as phase noise. Phase noise is the spectral density of the jitter versus offset frequency. It becomes an important parameter when designing phase-locked loops and clock recovery circuits. Electronics noise in the fiber optic link causes random jitter.

Figure 13.4: Fiber Optic Receiver — Noise to Jitter Conversion

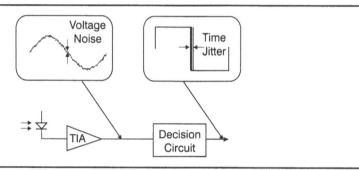

Primary contributors of noise in most modern fiber optic links include the transmitter light emitter (LED or laser diode) and the receiver photodiode preamplifier. Other elements contribute but are usually minor factors in modern fiber optic links. At low optical losses (i.e., high optical power at the receiver input) the transmitter is generally the limiting factor. At high optical losses (i.e., low optical power at the receiver input) the receiver noise usually dominates. Figure 13.5 shows the typical random jitter performance of a high-speed fiber optic link.

Figure 13.5: Random Jitter versus Optical Loss

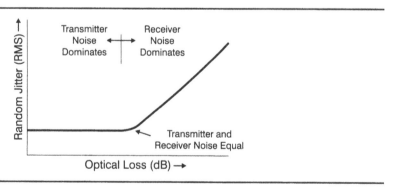

The next type of distortion is data dependent distortion or data dependent jitter. Data dependent jitter (DDJ) is also known as intersymbol interference. DDJ results from a frequency dependent amplitude distortion of the data stream. A pseudorandom NRZ data stream has a frequency spectrum that is a sin(x)/x function. Figure 13.6 illustrates the sin(x)/x function.

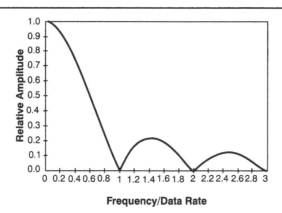

To achieve the most consistent data transmission the link must transmit the portion of the spectrum that contains most of the energy with minimal distortion. If the data link has non-flat frequency response in this critical region, intersymbol interference results.

Separate tests can determine random jitter and data dependent jitter independently. Random jitter can be tested using a "101010..." test pattern. Data dependent jitter can be tested by constructing a pattern that contains two distinct frequency components. For instance a pattern such as "111110000010" exercises the link at its maximum toggle rate (the last two bits) and at a rate that is 20% of the maximum rate (the first 10 bits). This will show any DDJ distortion very clearly.

The last type of distortion is duty cycle or pulse width distortion. This systematic distortion arises due to asymmetrical rise and fall times in some elements in the system. For instance, if a component has faster rise times than fall times, it will consistently make pulses too wide and vice versa.

COMMON TRANSMISSION STANDARDS

The variety of data links reflects a wide range of applications, data rates, and formats. Figure 13.7 shows many of the transmission standards now in use.

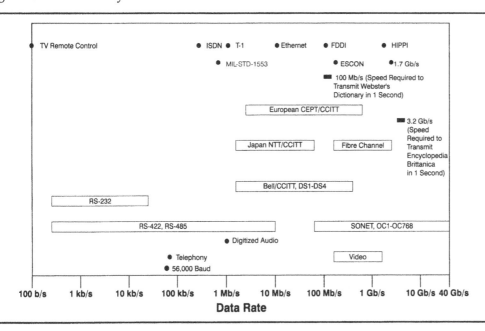

Figure 13.7:
Worldwide
Communication
Standards

The TV remote control reference point at 100 b/s is a free-space optical link that connects the television to the remote control. The other extreme, future telecom, indicates the movement of tomorrow's telecommunication links. Standards such as 56,000 baud usually connect computer modems to the phone line. The transmission standard labeled digitized audio refers to compact disk or CD audio that has a data rate of 1,411 kb/s. Data rates from RS-232 to RS-422/485 refer to short-haul data rates, while European CEPT/CCITT, Japan NTT, CCITT, BELL/CCITT, DS1-DS4 and SONET, OC-1 to OC-768 relate to current telecommunications links. MIL-STD-1553 is the military standard used in military craft. The transmission standard ESCON is gaining popularity as a means to connect computers to peripherals and computers to computers.

All transmission standards are arranged from lowest (100 b/s) to highest (40 Gb/s). A position on the chart around 100 Mb/s indicates the transmission speed required to send Webster's dictionary in one second. Another point of reference at 3.2 Gb/s indicates the speed required to transmit the Encyclopedia Britannica in one second.

HIGH-SPEED DATA TRANSMISSION

Data rate separates high-speed data links from other data links. This value indicates the number of data bits transmitted per second (bps or b/s). High-speed data rates usually include data rates over 100 Mb/s. In many ways, fiber can achieve data rates copper coax cable cannot. Large bandwidth, low attenuation, long distance capability, EMI/RFI, and lightning immunity exemplify key attributes that make fiber very advantageous for high-speed data transmission. However, choosing the components to develop a high-speed circuit requires special consideration because high bit data rates place demands on system components that low bit data rates do not. Frequency range, frequency-dependent components, parasitic components (such as capacitors), delay times must be given additional consideration when designing a high-speed data link.

The rapid growth of high-speed data links in network applications prompted the development of many protocols such as Fibre Channel, FDDI, SONET, and ATM (see Figure 13.7). Local area networks (LAN), wide area networks (WAN), and point-to-point communications may each use protocols uniquely suited to their requirements. However, connections to LAN's, WAN's, and point-to-point links further complicate matters. Storage area networks (SAN's), a variation on LAN's, typically connect a computer or group of computers to a number of high-capacity storage devices, often referred to as a RAID (redundant array of independent disks).

Figure 13.8:
Ethernet Applications for LAN, WAN and SAN

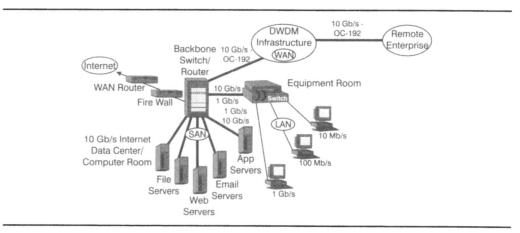

Most available fiber optic data link designs handle a single data protocol. However, recent developments include multi-protocol links capable of meeting requirements of several data protocols simultaneously. The multi-protocol link designer must meet both physical and performance expectations for all of the protocols handled by that link. At the physical layer, the requirements for many of the protocols are very similar. The optical transmitter must take an incoming electrical signal and convert it to an optical signal suitable for transmission over fiber. Conversely, the receiver must take an incoming optical signal and convert it to an electrical signal. Most protocols also require a method to disable the transmitter and require a loss-of-signal flag on the receiver.

On the other hand, challenging factors hinder the development of a multi-protocol high-speed data link. The optical dynamic range or optical loss range and receiver sensitivity vary from protocol to protocol. Optical dynamic range describes the range of optical input powers for which the receiver will deliver acceptable performance. Receiver sensitivity is the minimum average optical input power to the receiver for which it will deliver acceptable performance. Bandwidth requirements and fiber compatibility are also protocol-specific issues. The link bandwidth defines the maximum rate at which the link can transfer data. Maximum data rate in bits per second defines a data link. When designing a data link, one must know how much bandwidth in Hertz (Hz) is required to transfer a given data rate. Generally, for an NRZ data signal, the required bandwidth in Hz is 56% of the maximum data rate. This formula is given in Equation 13.6.

Eq. 13.6 $$BW\ (Hz) = 0.56 \cdot Data\ Rate\ (b/s)$$

For example, a data rate of 1 Gb/s would require about 560 MHz of bandwidth. Considering all these factors, to handle more than one data protocol, a multi-protocol high-speed data link must meet the following criteria:

- The receiver's sensitivity must be low enough to handle the lowest required sensitivity of the protocols in use.
- The receiver's optical dynamic range must be wide enough to handle the widest requirement.
- The transmitter's optical output power must be set such that the link can handle all loss budget requirements.
- The link bandwidth must be such that the highest data rates can be transferred while maintaining all specified levels of performance.

Such a link was developed by Force, Incorporated for use in both FDDI and fibre channel networks. These two important high-speed protocols bear further discussion.

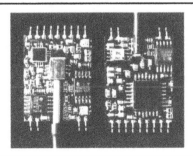

Figure 13.9: Multi-protocol High-speed Data Link

(Photo by Force, Inc.)

FDDI

The fiber distributed data interface standard was first introduced in the 1980's as the next step above ethernet and token-ring networks. The topology is a token-passing ring topology that uses two counter-rotating rings. The primary ring carries the information while the secondary ring functions as a backup in case of a component or other failure in the primary ring. Figure 13.10 illustrates this topology.

Figure 13.10: FDDI Dual Ring Topology

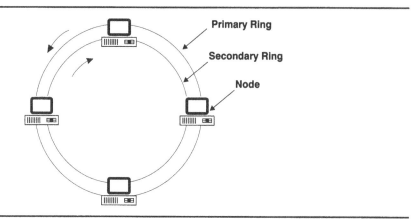

Data transmission is 100 Mb/s, but the 4B5B transmission code used in FDDI adds one extra bit for every four bits of data, making the actual data rate 125 Mb/s. The data is transferred between nodes which include a transmitter, a receiver, an electronic interface to the terminal, and an optional optical bypass switch. The transmitter/receiver pair act as a repeater. The receiver detects and amplifies the signal, then transmits it to the next node. The dual ring topology makes FDDI networks highly reliable. In the event of failure such as a cut cable or inoperable node, the network bypasses the problem by converting to a single ring topology until repairs are made. Many intelligent traffic systems (ITS) use a dual ring topology, usually based on RS-232 data rates, in order to improve reliability and survivability in case a fiber optic cable breaks or a unit fails.

FIBRE CHANNEL

The fibre channel high-speed network protocol operates at an average data transfer rate of 132 Mb/s to 1062.5 Mb/s. Initially developed to augment the small computer system interface (SCSI) protocol which transfers data at 20 Mb/s, the advance of such applications as multimedia, image processing, supercomputer modeling, and transaction processing to and from mass storage led to the need for a much faster transfer protocol. With a minimum data rate of 132 Mb/s, a maximum data rate of 1062.5 Mb/s, and enormous bandwidth, fibre channel meets these high-speed transfer requirements. In fact, work is being done to extend the data rates to 10 Gb/s and beyond.

Fibre channel topologies may be point-to-point or switched. As a point-to-point topology, the protocol has many attributes of a channel. It establishes dedicated, point-to-point connections between devices rather than using a shared medium. This allows the network to provide the full bandwidth to each connection. It is also hardware-intensive, relying on a frame header to trigger the routing of arriving data to the proper buffer. The switch topology connects a number of different devices where the switch acts as a central box that manages multiple point-to-point connections. In this topology, fibre channel has many of the attributes of a network. This bidirectional systems decouples the data encoding from the physical media and enables a high degree of scalability. Logical constructs replace physical control signals, eliminating the need for a channel control bus.

Fibre channel offers the ability to transport other network protocols such as asynchronous transfer mode (ATM), FDDI, ethernet, SCSI, and high performance parallel interface (HIPPI) over a single medium with the same hardware connection. All networks and protocols use buffers to hold the data being sent or received, giving fiber channel this advantage. Fibre channel uses 8B/10B encoding to provide the means to transfer this data between the sending buffer at the source and the receiving buffer at the destination without concern for the data format.

OC-192

Up to OC-48 data rates (2.5 Gb/s), the optical transmitter often used a directly modulated laser. At OC-192 data rates (10 Gb/s) and above, direct modulation of the laser becomes impractical because of the wavelength chirp that occurs when turning the laser on and off. This chirp causes a great deal of dispersion which severely limits the maximum transmission distance. In order to avoid laser chirp, the laser operates in a continuous mode, using an external modulator (see Chapter 7) to turn the light on and off before it enters the fiber.

OC-768/OC-3072

The telecommunications industry must next make the leap to OC-768, or 40 Gb/s. However, achieving these rates means overcoming significant problems, especially polarization mode dispersion (PMD). PMD compensation creates a challenge because unlike dispersion and nonlinearities, it changes over time. Therefore, developing a method of measuring and controlling PMD in real time or developing new data coding methods would help solve the problem. Pushing the envelope even further to OC-3072, or 160 Gb/s, will no doubt mean overcoming a whole new set of problems.

CONVERGENCE

Interestingly, a wide range of standards developed for very different purposes have begun to converge at a data rate near 10 Gb/s. Figure 13.11 illustrates the convergence of these and other data protocols as they have developed over the last decade.

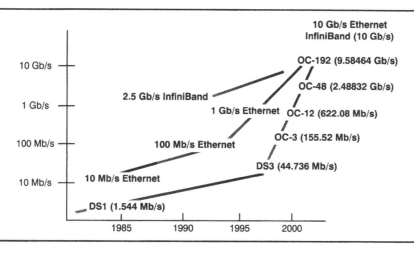

Figure 13.11: Convergence of 10 Gb/s Data Transmission Over Time

Chapter Summary

• When using fiber for data transmission, factors such as operating wavelength, optical loss budget, fiber type, fiber size, and optical connector type must be decided.

• Data transmission over fiber typically uses baseband transmission.

• Data scrambling takes an unknown, unpredictable data stream and converts it into a data stream with a predictable transition density.

- A few common data interfaces are TTL, CMOS, RS-485, ECL, and PECL.
- The key parameters associated with fiber optic data are data rate, bit error rate (BER), eye diagram parameters, jitter, and protocol.
- The required BER depends on the application and, to some extent, the data rate.
- The eye diagram is a powerful tool for analyzing the overall quality of a fiber optic data link.
- Although digital signals transmit data as a series of ones and zeroes, distortion can still occur.
- Types of digital distortion include random jitter, data dependent jitter, and duty cycle or pulse width distortion.
- Sources of noise in digital data systems may be the transmitter light emitter or the receiver photodiode preamplifier.
- Fiber optic data links have transmission standards that range from DC to 40 Gb/s.
- High-speed data links are an integral component in the rapidly growing areas of high-speed communications and high-speed networks, with standards such as fibre channel, FDDI, SONET, and ATM.
- Optical dynamic range describes the range of optical input powers for which the receiver will deliver acceptable performance.
- Sensitivity is the minimum average optical power to the receiver for which it will be able to deliver acceptable performance.
- Because of their dual ring topology, fiber distributed data interface (FDDI) networks are highly reliable.
- Fibre channel operates at an average data transfer rate of 132 Mb/s to 1062.5 Mb/s.
- One advantage of fibre channel is its ability to transport other network protocols.
- OC-192, OC-768, and OC-3072 use data rates up to 160 Gb/s to transmit data.
- Evolving fiber optic data rates have begun to converge around 10 Gb/s.

Selected References and Additional Reading

Bazaar, Charles. "Fibre Channel: The Future of High-Speed Connectivity." *Fiberoptic Product News.* May 1995: 34-35.

DiMinico, Chris. "10 Gigabit Ethernet — Convergence of LAN and WAN." BICSI 2000 Proceedings, Nashville, TN: August 2000, 1-44.

Hentschel, Christian. *Fiber Optics Handbook.* 2nd edition. Germany: Hewlett-Packard Company, 1988.

Kuecken, John A. *Fiberoptics: A Revolution in Communications.* Blue Ridge Summit, PA: Tab Professional and Reference Books, 1987.

Newell, Wade S. "Multi-Protocol High-Performance Serial Digital Fiber Optic Data Links," from *Digital Communications Design Conference, Day 3,* Joseph F. Havel, ed. Proceedings of Design SuperCon '95, Santa Clara, CA, Feb. 28-Mar. 2, 1995.

Shoemake, Mike and Lynn Woods. 1995. "FDDI in the" [sic]. *Fiberoptic Product News.* August 1995: 3-5.

Sterling, Donald J. *Amp Technician's Guide to Fiber Optics,* 2nd Edition. New York: Delmar Publishers, 1993.

PUSHING FIBER TO THE LIMITS

14

Now that we have completed the discussion of today's fiber optic system elements, the focus will shift to the fiber optic systems of tomorrow. The future of fiber optic advances lies in increased capacity for every aspect of the fiber optic system. However, the need to increase fiber capacity leads to the primary field of research for fiber optics.

TRENDS

Many of the developments discussed in this chapter originally evolved from other technologies such as wireless technology. The use of ever more exotic coding schemes and techniques, such as forward error correction (FEC), indicate that current fiber technology has yielded all of the "easy" bandwidth. These advanced techniques potentially increase the capacity of a single fiber from roughly 1 terabit capacity today to perhaps 10 terabits per second in the next five years. How high can fiber go? Years ago experts said of dial up computer modems that 28.8 kBaud was the absolute data rate limit. A few months later, designers announced the development of 56 kBaud modems. Fiber's absolute limit may or may not exceed 10 terabits per second; that goal seems achievable, but it will be extraordinarily complex and expensive. Figure 14.1 illustrates the potential capacity for the growth of fiber optic transmission rates.

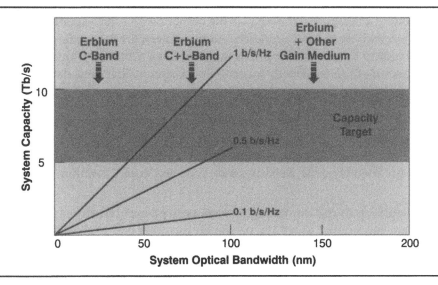

Figure 14.1: Potential System Capacity

Some estimates of fiber's ultimate capacity range from 25 Tb/s to as high as 100 Tb/s. As the bandwidth per fiber goes higher, the next concern regards the supportable distance. If the maximum distance reaches only a few kilometers, the ability to transmit 100 Tb/s on a single fiber holds little interest. Then there is the age-old concern of putting too many eggs in one basket. A network using only one fiber to carry a large percentage of the total traffic lacks the necessary security and reliability.

LIMITS TO FIBER'S GROWTH

Four key factors limit the maximum achievable data rate of an optical fiber: the optical amplifier bandwidth, polarization mode dispersion (PDM), crosstalk, and nonlinearities. Surprisingly, the list does not include factors such as optical attenuation and dispersion. Optical amplification and dispersion management overcome these potential limits. Optical amplifier bandwidth determines the available amount of a fiber's optical spectrum, and a wide bandwidth optical amplifier offers a greater usable area. High data rates are subject to distortion from PMD, a reality that cannot yet be easily and reliably overcome. In a system transmitting hundreds of wavelengths on a single fiber, crosstalk comes into play. The reality of imperfect filtering allows some of the undesired channels to leak into each active channel. Finally, fiber nonlinearities increase as optical power and fiber length increase and as optical channel spacing becomes closer. All of these factors set limits to the maximum data rate.

SPECTRAL EFFICIENCY

Spectral efficiency refers to the number of data bits per second that can be squeezed into one Hertz of bandwidth. As fiber pushes towards its data transmission limits, it adopts some of the tricks and techniques previously used on non-fiber communication channels. The 56 kBaud modem found in many home computers offers a good example. This device has a very high spectral efficiency, over 18/bits/second/Hz, meaning the device squeezes 56,000 bits per second through 3,000 Hertz of bandwidth. By comparison, a typical fiber optic application transmitting 10 gigabits per second of data transmits in a 50 GHz wide channel, has a spectral efficiency of only 0.2 bits/second/Hz. Researchers have developed exotic data coding schemes that improve on the number, with some experiments reporting spectral efficiencies close to 1 bit/second/Hz. Additionally, some data coding schemes carry an added benefit of better tolerating fiber nonlinearities. Obviously, any gains made in improving spectral efficiency directly increase the maximum fiber optic data rate.

FORWARD ERROR CORRECTION

Forward Error Correction (FEC) has been used for a number of years in submarine fiber optic systems. This technique that allows for near perfect data transmission accuracy even when faced with a noisy transmission channel. A number of FEC algorithms are being used including Hamming code, Reed-Solomon code and Bose-Chandhuri-Hocquenghem code.

By way of example, consider the operation of your cell phone just before your signal fades away. Let's say you wanted to tell the person on the other end of the call a sequence of numbers, and it was urgent that they hear the numbers accurately. Assume the list of numbers that you wish to transmit is 7, 3, 8, 10, 12, and 21. You could repeat the list of numbers twice and have the person at the receive end write down each list and compare them. If both received lists match, the data transmission is probably OK. Still, this requires you send all data twice, cutting the throughput of the system in half. Furthermore, if the lists do not match, there is no way to know which list is correct. Using this most basic method, to ensure that you could verify good transmission and correct some errors, you'd have to send the list three times and verify that two out of the three transmissions matched for each number.

A second method might work as follows: the sender first announces that they will send six numbers then resend the six numbers followed by the sum total of the six numbers. The transmission sequence would read 6, 7, 3, 8, 10, 12, 21, and 61. The person at the receive end would verify all the numbers and then ensure that the number at the end of the sequence equals the sum of the transmitted numbers. This system requires the transmission of less data, but any incorrectly received numbers will cause an incorrect value in the check sum number at the end of the transmission.

In a most basic way, these methods illustrate examples of error detecting codes. They determine transmission accuracy, but they do not correct the errors. The term "forward" in FEC implies that the error correction is accomplished by some information transmitted along with the data transmission. Error correcting codes are considerably more complex than error detecting codes and are ubiquitous in nearly every modern communication application. Error correcting codes even find widespread use in CD and DVD players.

In order to present an example of an error correcting code, we need to introduce and define two terms: binary and parity. In the prior examples of error detecting codes, we used numbers like 7, 3, 8, etc. These are base ten numbers, the 0-9 counting system familiar to us in everyday life. Binary numbers are base two numbers. They can only have two possible values, 0 or 1. Nearly all communication and computer systems use binary numbers.

Parity in binary communication systems indicates whether the number of 1's in a transmission is even or odd. If the number of 1's is an even number, then parity is said to be even and conversely for odd parity.

The Hamming Code

Consider a message having four data bits (D) which is to be transmitted as a 7-bit codeword by adding three error control bits. This would be called a (7, 4) code, meaning that the total codeword length equals seven bits, but only four of those bits contain data. The three additional bits are three even parity bits (P), where the parity of each is computed on different subsets of the message bits as shown below.

7	6	5	4	3	2	1	
D	D	D	P	D	P	P	7-Bit Codeword
D	,	D	,	D	,	P	Even Parity
D	D	,	,	D	P	,	Even Parity
D	D	D	P	,	,	,	Even Parity

For example, the message 1011 would be sent as 1010101, since:

7	6	5	4	3	2	1	
1	1	1	0	1	0	1	7-Bit Codeword
1	,	1	,	1	,	1	Even Parity
1	1	,	,	1	0	,	Even Parity
1	1	1	0	,	,	,	Even Parity

It may now be observed that if an error occurs in any of the seven bits, that error will affect different combinations of the three parity bits depending on the bit position. For example, suppose the above message, 1010101, is sent, and a single bit error occurs such that the codeword 1110101 is received:

Transmitted message   Received message

1010101   ------------>   1110101

Correcting the error requires examining which of the three parity bits was affected by the bad bit:

7	6	5	4	3	2	1	
1	1	1	0	1	0	1	7-Bit Codeword
1	،	1	،	1	،	1	Even Parity – OK
1	1	،	،	1	0	،	Even Parity – ERROR
1	1	1	0	،	،	،	Even Parity – ERROR

The pattern of the parity bit errors indicates in which bit in the codeword the error occurred, allowing correction of that bit. The value of this Hamming code can be summarized as follows:

1. Detection of two bit errors (assuming no correction is attempted);
2. Correction of single bit errors;
3. Cost of three bits added to a 4-bit message.

The ability to correct single bit errors costs less than sending the entire message twice. (Recall that simply sending a message twice accomplishes no error correction.) As the size of the codeword becomes very large, the extra burden of the error correcting bits becomes negligible. For instance, one possible Hamming code considered for fiber optic submarine transmission is the (18880, 18865) code. That means that the 18,880 bit codeword contains 18,865 actual data bits and 15 error correction bits. More robust forms of FEC may include more error correction bits that ensure the detection and correction of multiple bit errors in each codeword.

FEC on fiber optics uses a method similar to the Hamming code error correction given above. Typically, for an OC-192 system, FEC roughly adds 7% of overhead. The base data rate of 10 Gb/s increases to about 10.7 Gb/s. In other words, for every 1,000 bits of data transmitted, an additional transmitted 70 bits of FEC allow verification of the integrity of the received data and correct many of the errors that can occur during transmission.

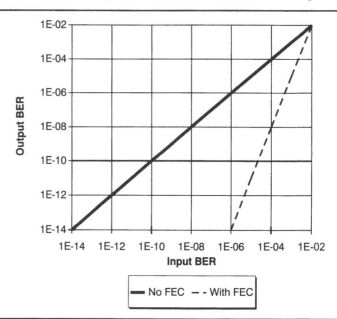

Figure 14.2: Effect of an FEC Algorithm on System BER

Figure 14.2 shows the effect of FEC on the system BER. As discussed in Chapter 13, BER measures the amount of received bits that have errors. A BER of 10^{-3} means that one out of every 1,000 bits is incorrect. The solid line shows the result if a given system does not have FEC. The input BER is a measure of errors that occur in the transmission channel. Since the system lacks FEC, any errors that occur during transmission appear at the system output. The dotted line shows what might happen with the addition of FEC to the system. Whereas before an input BER of 10^{-6} would yield an output BER of 10^{-6}, with FEC it now yields an output BER of 10^{-14}, a substantial improvement. Adding FEC to the system allows the designer to push distances and data rates farther than might be possible with any other technique. It also allows a system to have more operating margin.

1R/2R/3R REGENERATION

Because of the analog nature of transmission, even for digital data, optical signals undergo degradation as they travel through optical fiber due to dispersion, loss, crosstalk, and nonlinearities associated with fiber and optical components. Even after applying all of the compensating techniques, the signals require regeneration periodically. There are three different levels of regeneration that can be applied to a given signal. They include:

1R Regeneration: Amplification (Signal size increases but does not change shape.)

2R Regeneration: Amplification and Reshaping (Signal size increases in size, limited to give distinct "0" and "1" levels.)

3R Regeneration: Amplification, Reshaping and Re-timing (Signal size increases, limited to give distinct "0" and "1" levels, and the signal's timing information is restored.)

Figure 14.3 illustrates the difference in 1R, 2R, and 3R regeneration

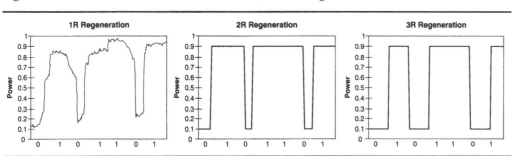

Figure 14.3: 1R/2R/3R Regeneration

Specially designed EDFA's or SOA's can perform 1R or 2R regeneration. Currently 3R regeneration requires converting the optical signal to electrical, performing the required clock extraction and retiming function, and then converting the signal back to optical. Research continues on all-optical components that could provide 3R regeneration, but these devices are not practical yet.

SOLITONS, RZ, AND NRZ TRANSMISSION

The discovery of the concept of solitons pre-dates fiber optics by more than 130 years. John Scott Russell discovered the concept of solitons in 1834 when he observed a wave propagate away from the bow of a canal barge without distortion. He followed that wave on horseback for several miles and found that it maintained it's shape over that extraordinary distance.

In fiber optics, solitons are very narrow pulses of light that have a very specific shape. The soliton travels down the fiber and interacts with dispersion in the fiber and nonlinearities in the fiber, the initial pulse shape is maintained. This differs fundamentally from coding schemes such as NRZ (Non-Return to Zero). As NRZ coded data encounters dispersion and nonlinearities in the fiber, the pulse shapes steadily degrades until they eventually become unusable. The ability of soliton pulses to travel on the fiber and maintain its launch wave shape makes solitons an attractive choice for very long distance, high data rate fiber optic transmission systems.

Figure 14.4: Typical Soliton Pulse

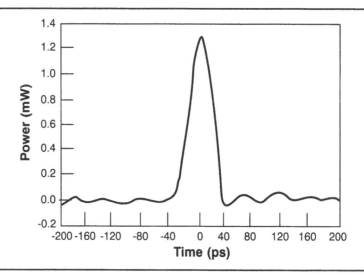

Figure 14.4 shows the shape of a soliton pulse. RZ (return-to-zero) pulses closely resemble solitons. RZ pulses are very narrow pulses of light; however, they lack the precise control of the pulse shape required with solitons, making them easier to generate. Unfortunately, RZ pulses accumulate distortion as they encounter dispersion and nonlinearity in the fiber. So while they are incapable of the same long distance transmission as solitons, their ease of creation and similarity in performance make them a viable alternative.

Let's take a close look at NRZ and RZ formats to better understand the distinction. Figure 14.5 shows typical NRZ data format and Figure 14.6 shows a typical RZ data format. The sequence shown in both cases is for an eight-bit data sequence (10011101) starting at time "1" and ending at time "9." The NRZ data format uses a steady light level to indicate whether a zero or a one is being transmitted. If a zero is being transmitted, no light (or very little light) is sent. If a one is being transmitted, a high light level is sent. The RZ data format also sends no light when a zero is being transmitted. When a one is to be sent, a very narrow pulse of a high light level is sent.

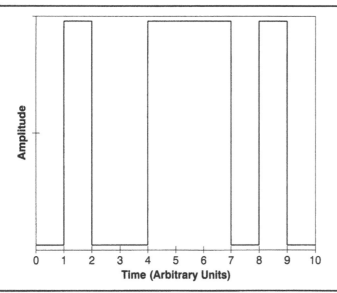

Figure 14.5: NRZ Data Format for 8-Bit Data Sequence (10011101) starting at time "1" and ending at time "9"

It can be said that the RZ data format only sends the rising edge information from NRZ data format. RZ data transmission has the advantage of being less susceptible to various distortions and nonlinearities that occur in optical fibers. On the other hand it is more complex to generate and detect than the NRZ data format.

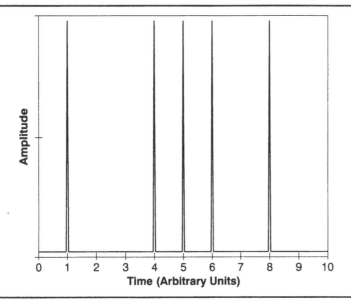

Figure 14.6: RZ Data Format for 8-Bit Data Sequence (10011101) starting at time "1" and ending at time "9"

Dispersion Managed Soliton Transmission

Dispersion management and solitons are essential technologies for anyone trying to develop a long-haul, DWDM fiber optic system operating at OC-192 or higher rates. Dispersion managed soliton transmission provides advantages over normal methods of dense wavelength-division multiplexing. The balanced use of nonlinearity and dispersion enables long-haul signal propagation without distortion. Higher spectral efficiency and polarization mode dispersion (PMD) tolerance is also achieved.

Dispersion management and solitons become especially important when the target fiber is NDSF. NDSF fiber without dispersion management or soliton technology could only support the transmission of OC-192 to a distance of 4 km, according to one source. When these technologies are optimally applied, the distance can be stretched to more than 1,000 km.

PMD COMPENSATION

Chapter 3 discussed causes and effects of polarization mode dispersion (PMD) in single-mode fiber. In long-haul optical transmission systems at 10 Gb/s, chromatic dispersion can be reduced by making use of dispersion compensated fiber (DCF). PMD compensation presents greater challenges, especially since it continually changes over time and can be difficult to track. Figure 14.7 illustrates the need to compensate for PMD by demonstrating its effects on three of the most popular telecommunications systems.

Figure 14.7: PMD-limited Distance of OC-48, OC-192 and OC-768

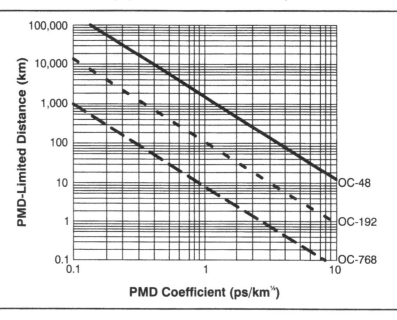

There are a number of techniques that can accomplish PMD compensation. Today, the most common approach uses PMD tolerant data coding formats. Figure 14.8 shows another approach to PMD compensation, the incorporation of a variable birefringence device. As data rates move beyond 10 Gb/s, researchers will need to develop new adaptive PMD compensation techniques.

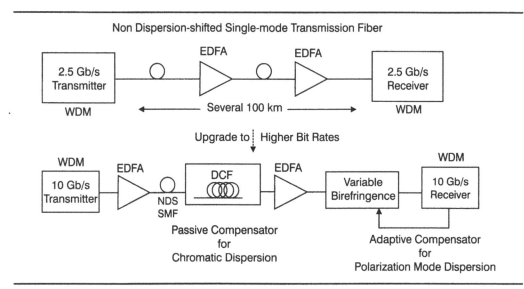

ULTRA LONG-HAUL TELECOMMUNICATIONS

Before WDM and DWDM became popular technologies, the telecommunications industry addressed its need for more bandwidth by developing a technique know as time-division multiplexing (TDM). This technique allowed several low speed data signals to be multiplexed into a single high speed data stream by placing each type of data signal in a specific sequence related to time during the transmission. While this technique gave the telecommunications industry some much needed bandwidth, it is still limited compared to DWDM's ability to increase fiber capacity. Figure 14.9 illustrates this difference.

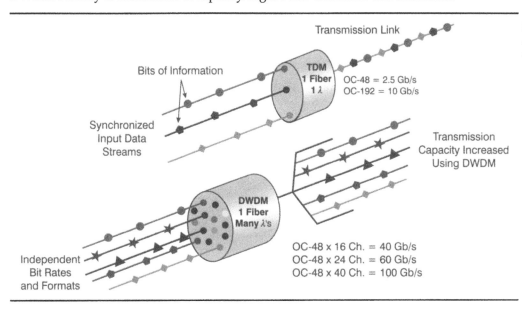

Figure 14.9: TDM versus DWDM Fiber Transmission Capacity

Long-Haul, OC-192, Multichannel DWDM Transmission

In order to achieve the goals of long-haul transmission, a multichannel DWDM transmission system must have several features:

1. Long Distance Transmission — The ability to transport the optical signals 500 km or more using only optical amplification.

2. OC-192 - Transmission at 10 Gb/s or higher.

3. Multichannel DWDM — The ability to send a eight or more optical wavelengths per fiber.

4. The ability to accomplish items 1 through 3 on NDSF single-mode fiber.

These basic elements require a very sophisticated system. At 1550 nm, it is only possible to transmit an OC-192 signal about 4 km on NDSF without special coding (such as RZ or soliton) and/or dispersion management. A system would have to employ some combination of special coding and highly accurate dispersion control to meet the requirements. The first three elements directly affect the data capacity of a given optical fiber. The last element is critical to the rapid acceptance of such a system since NDSF was the most common type of fiber installed in the late 1990's.

THE RECIRCULATING LOOP

For now, most of the leading edge experiments on fiber transmission capacity are performed in a laboratory environment using a recirculating loop as shown in Figure 14.10. In this setup, the light is launched into a loop that contains fiber, compensating elements, and optical amplifiers. The light travels around the loop multiple times to simulate a long distance. The signal is then extracted at the appropriate time and analyzed. This allows very long distance transmission (often >10,000 km) to be simulated with a modest amount of fiber.

Figure 14.10: Typical Recirculating Loop Test Setup

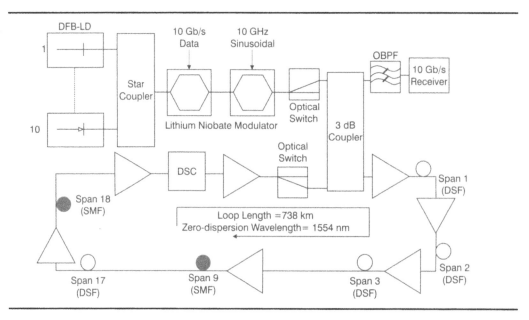

Chapter Summary

- Expanding the transmission capacity represents the greatest trend in fiber optic development today.

- Four key factors limit the maximum achievable data rate of an optical fiber: the optical amplifier bandwidth, polarization mode dispersion (PDM), crosstalk, and nonlinearities.

- Spectral efficiency refers to the number of data bits per second that can be squeezed into one Hertz of bandwidth.

- Forward error correction (FEC) allows for near perfect data transmission accuracy even when faced with a noisy transmission channel.

- FEC algorithms in use today include the Hamming code, Reed-Solomon code, and Bose-Chandhuri-Hocquenghem code.

- Three different levels of signal regeneration include 1R, 2R, and 3R regeneration.

- Solitons are very narrow pulses of light that have a very specific shape.

- As non return-to-zero (NRZ) coded data encounters dispersion and nonlinearities in the fiber, the pulse shapes steadily degrade until they eventually become unusable.

- Return-to-zero (RZ) pulses, close cousins to solitons, have very narrow pulse shapes, but they lack the precise control of the pulse shape required with solitons, making them easier to generate.

- Dispersion managed soliton transmission provides advantages over normal methods of dense wavelength-division multiplexing because the balanced use of nonlinearity and dispersion enables long-haul signal propagation without distortion, higher spectral efficiency, and PMD tolerance.

- Polarization mode dispersion compensation allows fiber to meet its potential transmission bandwidth, but the technology presents challenges because PMD continually changes over time and can be difficult to track.

- In order to achieve the goals of long-haul transmission, a multichannel DWDM transmission system must have the ability to transport eight or more 10 Gb/s (or higher) optical signals for 500 km on non dispersion-shifted single-mode fiber using only optical amplification.

- Most of the leading edge experiments on fiber transmission capacity are performed in a laboratory environment using a recirculating loop wherein light is launched into a loop that contains fiber, compensating elements, and optical amplifiers. The light travels around the loop multiple times to simulate a long distance, allowing very long distance transmission (often >10,000 km) to be simulated with a modest amount of fiber.

Selected References and Additional Reading

Bazaar, Charles. "Fibre Channel: The Future of High-Speed Connectivity." *Fiberoptic Product News.* May 1995: 34-35.

Cowper, Rich, Maurice O'Sullivan, Gen Ribakovs, and Kim Roberts. "Perspectives on the Evolution of Next Generation Higher Capacity Systems." Nortel Networks, 1999: 1-38.

Hentschel, Christian. *Fiber Optics Handbook.* 2nd edition. Germany: Hewlett-Packard Company, 1988.

Hidenori Taga, Kaoru Imai, Noriyuki Takeda, Masatoshi Suzuki, Shu Yamamoto, and Shigeyuki Akiba. "100 WDM x 10 Gbit/s Long-distance Transmission Experiment Using a Dispersion Slope Compensator and Non-Soliton RZ Pulse. "*Optical Society of America, Trends in Optics and Photonics*, TOPS vol. XVI, July 21-23, 1997: 382-385.

Kuecken, John A. *Fiberoptics: A Revolution in Communications.* Blue Ridge Summit, PA: Tab Professional and Reference Books, 1987.

Lagasse, P., P. Demeester, A. Ackert, W. Van Parys, B. Van Caehegem (IMEC), M. O'Mahoney, A. Tzanakaki (UoE), K. Stubkjaer (DTU), J. Benoit (ENST). "Roadmap Towards the Optical Communications Age." *A European view by the Horizon project and the ACTS Photonic Domain.* November 1999.

Newell, Wade S. "Multi-Protocol High-Performance Serial Digital Fiber Optic Data Links," from *Digital Communications Design Conference, Day 3*, Joseph F. Havel, ed. Proceedings of Design SuperCon '95, Santa Clara, CA, Feb. 28-Mar. 2, 1995.

Shoemake, Mike and Lynn Woods. 1995. "FDDI in the" [sic]. *Fiberoptic Product News.* August 1995: 3-5.

Sterling, Donald J. *Amp Technician's Guide to Fiber Optics*, 2nd Edition. New York: Delmar Publishers, 1993.

TESTING & MEASUREMENT TECHNIQUES

15

Standardized measurement and testing techniques are important to precisely define the parameters under which a fiber optic system or any of its components will function. In evaluating the performance of a system, all the components that comprise the system must be considered. Techniques for measuring fiber optic systems fall into two major categories: functional testing and performance testing. Functional testing involves determining whether a component of the fiber optic system is operating. For example, testing fiber continuity with the use of an optical time-domain reflectometer represents functional testing. Performance testing involves how the system performs relative to the expected performance determined by the optical link loss budget.

FIBER OPTIC TEST EQUIPMENT

Test equipment for installation and troubleshooting fulfill an important part of a fiber optic system. Several important types of test equipment in use now include:

Optical Power Meter: This instrument measures the amount of optical power in a fiber. Most models handle several wavelengths and provide relative (dB) as well as absolute (dB or Watts) measurements. Multiple adapters are usually required to deal with different optical connector types.

Optical Light Source: An optical light source injects a stable test light signal into a fiber. Most models offer continuous wave modes. Some offer modulation modes as well. Typical test modulation frequencies are 270 Hz, 1 kHz and 2 kHz.

Optical Loss Meter: This instrument combines an optical power meter and an optical light source into a single instrument. It is also called an optical loss test set (OLTS).

Fiber Identifier: A fiber identifier traces (locates) the path of a fiber. This is accomplished by injecting a tracer signal into the fiber and using a tapping device on the fiber. The tapping device causes a small amount of light to be detected as it leaks out of the fiber. The light that leaks out is captured and processed to determine if it matches the tracer signal.

Talk Set: Talk sets are used to coordinate maintenance activities over an optical fiber. Often, optical fibers are underground or in shielded rooms where walkie-talkies are unusable. In these cases, the fiber itself is the best means of communicating. Talk sets are available for operation on one fiber or two fibers and offer half-duplex or full-duplex communications.

Optical Time Domain Reflectometer (OTDR): OTDR's are one of the most useful installation and diagnostic tools available today. Figure 15.1 shows the typical output of a modern OTDR. In this example, the OTDR is examining two lengths of fiber that have been connected together. The wavelength being used is 1310 nm. The first length of fiber shows no anomalies in its 19.3784 km length. The average attenuation of that first length of fiber is 0.505 dB/km. At the end of the first length, a connector causes considerable backreflection (-13.72 dB) and introduces 1.404 dB of optical loss. The second length of fiber is 19.397 km long and has an average attenuation of 0.353 dB/km. The connector at the end of the second length of fiber also has a strong backreflection of -14.56 dB. The region after the second connector is noise.

Figure 15.1: Optical Time Domain Reflectometer

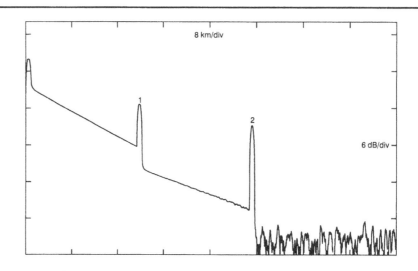

Table 15.1: OTDR Settings for Figure 15.1.

Wavelength	Pulse Width	Range	Pt. Space	R. Index	Backscatter
1310 nm	500 m	64 km	10.000 m	1.4660	-79.20 dB
Event No.	Location	Loss (dB)	Atten (dB)	Refl (dB)	
1	19.3784 km	1.404	0.505	-13.72	
2	38.7754 km	5.207	0.353	-14.56	

Optical Spectrum Analyzer: Optical spectrum analyzers determine the wavelength of light. This device determines the emission spectrum of a light source.

Fiber Optic Attenuator: A fiber optic attenuator, also called an optical attenuator, simulates the loss that would be caused by a long length of fiber. Typically, this device performs receiver testing. While an optical attenuator can simulate the optical loss of a long length of fiber, it cannot accurately simulate the dispersion that would be caused by a long length of fiber.

Figure 15.2: Variable Fiber Optic Attenuator

(Photo courtesy of PD-LD, Inc.)

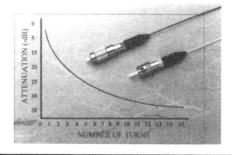

Backreflection Meter: Modern high bit rate digital fiber optic links and most laser-based analog links require very low backreflection to operate properly. This instrument quantifies the amount of backreflection in the fiber path.

Local Injector Detectors: These are used to measure and/or assist in the tuning of fiber splices. They do so by injecting and detecting light through the sides of the fiber, usually using sophisticated microbending techniques. For rotary splice tuning, the light is detected through the splice itself rather than the fiber.

MEASUREMENT TECHNIQUES FOR COMPONENTS

Fiber Measurements

Optical fiber measurements vary depending on whether the fiber is multimode or single-mode. Tests for multimode fiber measure attenuation, multimode dispersion, chromatic dispersion, numerical aperture, and core diameter. Important single-mode fiber tests measure attenuation, chromatic dispersion, cutoff wavelength, and spot size.

Attenuation is a critical fiber parameter. Varied mode propagation complicates attenuation testing. Attenuation is tested by exciting the fiber to equilibrium mode distribution (EMD). EMD represents the modal distribution through a long length of fiber. An optical source and a power meter provide the means for testing attenuation. The insertion loss for a short reference fiber is compared to the insertion loss of the longer length test fiber. This test is only useful if care is taken to accomplish the same coupling efficiency in both cases. By analyzing the fiber's backscatter signal with an optical time-domain reflectometer, the uniformity of the attenuation along the fiber length can be determined.

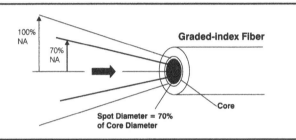

Figure 15.3: EMD — Excitation of Graded-index Fiber

Automated switching systems for testing components such as cable, connectors, or couplers are beneficial in reducing the cost of test equipment, because several devices can be tested simultaneously. Figure 15.4 demonstrates a setup to test cable using an automated switching system.

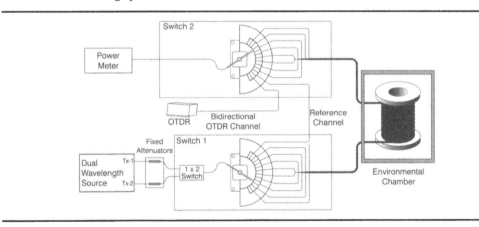

Figure 15.4: Cable Testing

Light Source Measurements

Both LED's and laser diodes, must be considered when measuring fiber optic systems. LED measurement tests include output power, modulation bandwidth, center wavelength, spectral width, source size, and far-field pattern. Tests for laser diodes include optical parameters, output power, modulation bandwidth, center wavelength/

number of modes, chirp/linewidth, and mode field of the Gaussian beam as well as electrical parameters such as threshold current, slope efficiency, forward voltage, monitor current, etc.

LED's exhibit a nearly linear dependence of the output power on the drive current. Power from the fiber pigtail can be measured by a power meter. Because of the wide spectral width of the emitted radiation, the accuracy of the measurement depends on the wavelength-dependence of the power meter's detector.

The modulation bandwidth measurement uses an LED intensity-modulated with a sweep-generator, and a PIN diode which reconverts the optical signal back to the electrical domain. The frequency response can be observed on an oscilloscope. Another approach uses a network analyzer with an optical-to-electrical conversion at the receiver end.

An optical spectrum analyzer measures center wavelength as well as spectral width. The result of this measurement will influence the pulse broadening of the fiber. The size of the radiating area as well as far-field should be measured directly at the LED chip (without the fiber). The size can be determined by analyzing the microscopic image (near-field), whereas the measurement of the angular power-distribution located some distance from the source will yield the far-field. Narrow near-fields and far-fields are necessary for high coupling efficiency to the fiber.

Output power can be measured with a variable current source and a power meter. For lasers, threshold current defines the onset of the stimulated emission.

Modulation bandwidth measurement techniques for lasers are identical to those described for LED's. However, because lasers operate at a much higher bandwidth, the measurement equipment must be able to read higher bandwidth. Center wavelength as well as the number of modes should be measured with an optical spectrum analyzer.

Chirp is the undesired wavelength shift caused by intensity modulation. Laser diodes that emit a single linewidth are affected by chirp. Interferometric methods are capable of measuring both chirp and linewidth.

The radiation characteristics of a laser diode can be approximated by an elliptic version of the Gaussian beam. A Gaussian beam has a finite beam width which smoothly transits into a light cone of fixed numerical aperture. It is rotationally symmetrical to the direction of propagation. The ellipticity is caused by the fact that the emitting area is a stripe rather than a circle. A far-field measurement (analysis for the power density at some distance from the radiating area) will deliver the parameters of the Gaussian beam.

Light Detector Measurements

Tests for PIN photodiodes and APD's include diameter, spectral responsivity, bandwidth, and dark current/NEP. APD's are additionally tested for multiplication factors and excess noise. Detectors perform two functions: signal detection in receivers and optical power measurement. When the detector's function is signal detection, the smallest possible diameter is the desirable measurement because the noise equivalent power (NEP) is proportional to the active diameter, and the bandwidth is inversely proportional to the active area. When the detector is acting as an optical power detector, the desirable diameter is the largest possible measurement because it increases power measurement accuracy.

For both PIN photodiodes and APD's, spectral responsivity is strongly dependent on wavelength. A wavelength-calibrated combination of a tungsten lamp and a tunable monochromator measure spectral responsivity. Ideal responsivity would be proportional to the wavelength; however, actual readings may vary considerably. The multiplication factor measured in APD's is a result of high voltage which leads to the multiplication of the number of generated carriers. This multiplication factor is measured the same as spectral responsivity.

Detector bandwidth can be measured by exciting the detector with a sinewave-modulated laser source. A network analyzer with an electrical-to-optical conversion of the generator signal can be used to perform the measurement. Noise equivalent power is critical because it influences the achievable receiver sensitivity. Ideally, the NEP is proportional to the square root of the dark current. A more accurate measurement can be taken using a spectrum analyzer. In APD's, the excess noise factor is an additional noise contribution caused by the multiplication process mentioned above. It too is measured with a spectrum analyzer.

Interconnection Loss Measurements

The ideal interconnection of one fiber to another would have two fibers that are optically and physically identical held by a connector or splice that squarely aligns them on their center axes. However, in the real world, system loss due to fiber interconnection is a factor. Insertion loss is the primary consideration for connector performance. There are three types of insertion loss: fiber-related loss, connector-related loss, and contributing system factors. Because of the discrepancy between insertion-loss testing and connector performance, users must understand the test methods used to measure insertion loss. The best test results are obtained when lengths of fibers are attached to the source and detector as permanent parts of the test setup. This avoids variations in results that are caused by source and detector interconnection losses from test to test.

Insertion loss tests will reduce the influence of fiber-related losses. A general test should be reproducible and provide applicable results. Most tests measure the output power (P_1) of a length of fiber. The fiber is then cut in the middle and terminated with a connector or splice. The output power (P_2) is measured again. Insertion loss is given by:

Eq. 15.1
$$\text{Loss (dB)} = 10 \cdot \log_{10}\left(\frac{P_1}{P_2}\right)$$

The length of fiber must be broken perfectly in the middle to produce an identical fiber on each side of the splice. This method purposely eliminates fiber-induced losses in order to evaluate connector performance independently of fiber-related variations. Three sets of launch conditions are of interest.

1. Short-launch, short-receive: Represented by short fibers with no mandrel wrap on the transmitting or receiving ends. Short-launch, short-receive conditions exhibit losses that increase with the slightest mechanical offset of the connection. Lateral misalignment is a critical parameter under these conditions.

2. Long-launch, short-receive. A mandrel wrap is on the transmitting end but not on the receiving end. This condition reduces the exit NA of the transmitting fiber and end-separation losses. Since all of the receiving core can be used, greater separation of the fibers can be tolerated.

3. Long-launch, long-receive. A mandrel wrap at both the transmit and receive ends creates the condition, showing a greater sensitivity to lateral misalignment than the other two conditions. Reducing the effective core area offset of both fibers increases loss significantly.

The insertion loss test assumes the use of two pieces of identical fiber. The use of two different types of fibers requires accounting for NA mismatch loss and diameter mismatch loss.

NA mismatch loss occurs when the numerical aperture of the transmitting fiber (t) is larger than that of the receiving fiber (r). Figure 15.5 illustrates NA mismatch loss.

Figure 15.5: NA Mismatch Loss

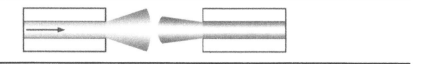

The calculated loss for numerical aperture mismatch is approximated by:

Eq. 15.2
$$Loss_{NA} = 10 \cdot log_{10}\left(\frac{NA_r}{NA_t}\right)$$

As illustrated in Figure 15.6, core diameter mismatch occurs when the core diameter of the transmitting fiber (t) is larger than the core diameter of the fiber at the receiving end (r). Cladding diameter mismatch is similar to core diameter mismatch loss except the cladding of the transmitting fiber differs in diameter from the cladding of the receiving fiber. Either mismatch prevents the cores from aligning.

Figure 15.6: Core-diameter Mismatch Loss

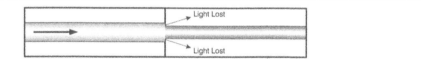

Both types of diameter mismatch loss are approximated by:

Eq. 15.3
$$Loss_{dia} = 10 \cdot log_{10}\left(\frac{dia_r}{dia_t}\right)$$

This equation is only accurate if all of the modes in the fiber are excited. When only low-order modes are excited, the loss is greatly reduced and may not be present at all.

Concentricity, also known as eccentricity, occurs because the core may not be perfectly centered in the cladding. Ellipticity or ovality describes the fact that the core or cladding may be elliptical rather than circular. The alignment of the two elliptical cores will vary depending on how the fibers are brought together. These forms of connector loss are illustrated in Figure 15.7.

Figure 15.7: Concentricity and Ellipticity

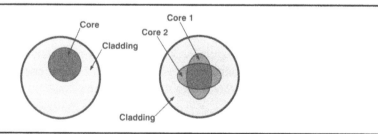

Connector-related loss can also result from the mechanical misalignment of the optical fiber cores. There are several types of misalignment loss: lateral displacement, angular misalignment, and end separation.

A connector should align the fibers on their center axes, but when one fiber's axis does not coincide with the other fiber's axis, lateral displacement occurs. A displacement of only 10% of the core axis diameter results in a loss of about 0.5 dB. The ends of mated fibers should be perpendicular to the fibers' axes and to each other. Failure to be perpendicular is called angular misalignment.

Some connectors hold the two fibers slightly apart to prevent the fibers from rubbing against each other and damaging their end polishes. Fresnel reflection loss or end separation loss is caused by the difference in the refractive indices of the two fibers and the air that fills the gap between the two fibers. Some connector manufacturers believe the use of index-matching gel in the gap reduces Fresnel reflection loss, but others do not recommend using index-matching gel. This gap may collect small flecks of abrasive contaminates that will damage the end finishes, and the addition of index-matching gel could compound this contamination.

In a single-mode interconnection with a flat end finish, Fresnel reflection loss can be as much as -11 dB, a level sufficient to disrupt the operation of most lasers. This loss can be reduced by rounding the fiber end of one fiber during polishing (called a PC or physical contact finish). While it would seem practical to use a flat finish and butt the ends, getting two perfectly smooth, flat finishes is nearly impossible. With a rounded finish, fibers always touch on a high point near the light-carrying fiber core (see Figure 15.8).

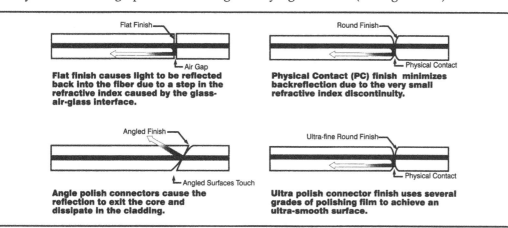

Figure 15.8: Fiber End Face Finishes

System-related factors in connector loss involve the launch and receive conditions. These conditions result from the mode distribution in the fibers. The performance of the connector depends on modal conditions and the connector's position in the system. Four different conditions exist:

1. Short launch, short receive.
2. Short launch, long receive.
3. Long launch, short receive.
4. Long launch, long receive.

These launch conditions must be controlled in order to provide repeatable measurements. Long-launch conditions are generally preferred. Long-launch or receive conditions mean that equilibrium mode distribution (EMD) exists in the fiber. The Electronic Industry Association (EIA) recommends a 70/70 launch: 70% of the fiber core diameter and 70% of the fiber NA should be filled. This recommendation corresponds to the EMD in a graded-index fiber. EMD can be reached three ways: by the optical approach (illustrated in Figure 15.3), filtering, or long fiber length. In general, connector losses under long-launch conditions range from 0.4-0.5 dB. Under short-launch conditions, losses are in the 1.3-1.4 dB range.

Splice loss testing for preterminated cable involves first measuring the power (P_1) transmitted through two separate fibers. Then the cable is joined in a splice bushing, and a new transmission power reading is made (P_2) through the combined cable.

The equation for calculating splice loss is written: ·

Eq. 15.4 $$\text{Splice Loss} = 10 \cdot \log_{10}\left(\frac{P_2}{P_1}\right)$$

Where:

P_1 = Power reading through the first cable.
P_2 = Power through the combined cables.

Systems Measurements

System measurements include fiber continuity, bit error rate (BER), sensitivity, and eye pattern. These are the functional tests. Fiber continuity is tested using an OTDR which locates breaks in the optical fiber. The number of false bits of information divided by the total number of bits that have been transmitted defines bit error rate. BER is tested by modulating the source with a well-defined, long, bit-sequence and comparing the received sequence to the transmitted one. Sensitivity is the lowest power level that leads to the desired BER. It can be tested with the BER test method; in this case, an optical attenuator is used to reduce the transmitted power until the error rate exceeds the predefined value. Eye patterns are viewed with an oscilloscope.

Chapter Summary

* Fiber optic systems may undergo functional testing or performance testing.
* Test equipment for fiber optic systems include an optical power meter, optical light source, optical loss meter, fiber identifier, talk set, optical time domain reflectometer, optical spectrum analyzer, optical attenuator, backreflection meter, and local injector detectors.
* Automated switching systems for testing components such as cable, connectors, or couplers are beneficial in keeping down the cost of test equipment.
* Center wavelength, as well as spectral width can be measured with an optical spectrum analyzer.
* Tests for PIN photodiodes and APD's include diameter, spectral responsivity, bandwidth, and dark current/NEP.
* Detector bandwidth can be measured by exciting the detector with a sinewave-modulated laser source.
* Concentricity, or eccentricity, occurs because the core may not be perfectly centered in the cladding.
* Ellipticity, or ovality, describes the fact that the core or cladding may be elliptical rather than circular.

Selected References and Additional Reading

—. *Designer's Guide to Fiber Optics*. Harrisburg, PA: AMP, Incorporated, 1982.

Hecht, Jeff. *Understanding Fiber Optics*. 2nd edition. Indianapolis, IN: Sams Publishing, 1993.

Hentschel, Christian. *Fiber Optics Handbook*. 2nd edition. Germany: Hewlett-Packard Company, 1988.

—. *Just the Facts*. New Jersey: Corning Incorporated, 1992.

Sterling, Donald J. *Amp Technician's Guide to Fiber Optics*, 2nd Edition. New York: Delmar Publishers, 1993.

FUTURE TRENDS

16

Predicting the future has always been a challenging and usually unsuccessful undertaking. In spite of the mountain of evidence that it cannot be done, we will attempt it anyway. First let us look at major trends that affect the fiber optics industry.

1. Increased fiber transparency
2. All-optical networks
3. Multi-terabit transmission
4. Competing technologies
5. Expansion into mass markets
6. Cost reductions
7. Miniaturization
8. New materials
9. Technology refinements
10. New technology

The order of the items is intentional. Depending on the specific product and market considered, the main thrusts in the fiber optics industry today cover in the first four to seven items. Most prognosticators of the future focus on number ten, new technology. Today, in the second age of fiber optics, the DWDM age, new technology is having a huge impact on the industry.

Expansion into mass markets and cost reduction go hand-in-hand. Fiber optics has always suffered from the chicken and egg syndrome. High volumes would create a price drop, but high volumes would not be realized until prices dropped. For example, many large, well-financed fiber optics companies went bankrupt in the early 1990's gambling that fiber-to-the-home (FTTH) would take off at any moment. It didn't, and so far it hasn't. It probably will, but exactly how and when remains the multi-billion dollar question. Let's look at each of the ten points listed in some detail to get a better picture of what is happening and what is likely to happen.

INCREASED FIBER TRANSPARENCY

Earlier, this book discussed the four windows used in fiber optic transmission and focused on the third window, the "C-Band" near 1.55 μm. The large "no mans land" between the second and third windows has been unused to date because of the large OH$^-$ peak near 1.38 μm. Modern fibers have been refined to the point that this peak can be all but eliminated. This opens the possibility of a fifth window stretching from about 1.35 μm to 1.53 μm. This would give a continuous wavelength band from 1.26 μm to 1.65 μm providing nearly 50 THz of bandwidth. An additional 190 THz of bandwidth lies between the first and second windows. This region would be more difficult to exploit since most of it lies below the cutoff wavelength for single-mode fibers. In order to exploit the 50 THz band that spans the second, third, fourth and fifth windows, super wideband optical amplifiers would have to be developed. Figure 16.1 below shows the first through fifth windows.

Figure 16.1: The Five Windows in Optical Fiber

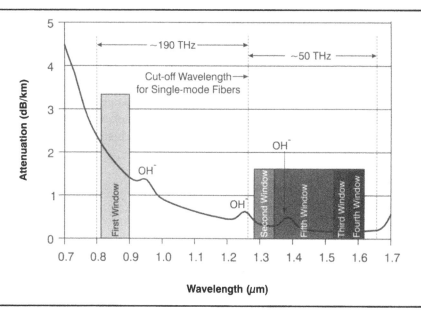

ALL-OPTICAL NETWORKS

All-optical networks are one of the hot topics in fiber optics. The basic strategy processes all signals in the optical domain. Currently, in most signal processing and switching, the optical signal is first converted to an electrical signal, processed, and switched electrically before conversion back to an optical signal. This is referred to as an O-E-O signal processing chain. Optical signals can handle data rates far faster than today's electronics, creating an interest in all-optical networks. If all processing and switching could be done optically, then it may not be necessary to replace electronics every time data rates increase. Currently, fiber optic transmitters and receivers can transmit only a single data rate, thus, they must be replaced when the data rate increases.

Many challenges complicate the creation of a true all-optical network. Functions such as switching and optical regeneration, at least 1R and 2R regeneration (e.g. an EDFA) have relatively straightforward approaches. Other functions such as 3R regeneration, reading headers on the optical signals, switching the optical signal on the fly based on the header content and real-time wavelength swapping, represent just a few of the challenges that need addressing in order to make all-optical networks a reality.

MULTI-TERABIT TRANSMISSION

The availability of more bandwidth drives the development for multi-terabit transmission. The one terabit per second barrier has been broken by using a combination of high-speed data transmission (10 Gb/s), and more than 100 DWDM channels. The data rate will, no doubt, continue to climb, and now researchers have set their sights on two terabits per second, five terabits per second, and more. The pressure to carry even more DWDM channels per fiber seems relentless. This speed comes at a high cost, and only can be justified on long-haul systems. The need to install new fiber drives up the cost, with fiber ranging in price from $50,000 to $500,000 per mile.

COMPETING TECHNOLOGIES

Twenty years ago, futurists predicted that the use of copper cable would be all but gone for most communication applications, a bold, but wrong, prediction. Copper is still here today and will be twenty years from now. In general, copper remains the first choice for short distances, and fiber remains the first choice for long distances. For most applications, there is a crossover distance at which the cost of a copper solution and a fiber solution are equal. This crossover distance is bandwidth dependent: the higher the bandwidth, the shorter the crossover distance. At distances shorter than the crossover distance, copper is cheaper and at distances greater than the crossover distance, fiber will yield lower costs. Ten years ago, this crossover distance was several kilometers. Today it may be as short as 100 meters for some applications.

Fiber's competition today will not be its only competition in the future. Five years ago, no one would consider sending an NTSC video signal over one kilometer of twisted pair cable. Today, it can be done, and that removed some of fiber's opportunities. Advances in compression technologies will also erode some fiber opportunities by decreasing the needed transmission bandwidth. Other improvements in competing technologies will continue to arise in the future, but the inherent advantages of fiber will continue to make it an essential technology.

EXPANSION INTO MASS MARKETS

DWDM and the Internet Demand

No one really predicted the enormous success of the Internet and its associated components, the world wide web and electronic mail. In just a few short years, the network of literally millions of computers moved from a mainly military and research-oriented system to a commerce-oriented venue that offers any type of information, image, or product imaginable. Fiber optics offered the only practical means of meeting the demand for this growth. Early on, the telecommunications industry installed a great deal of fiber in the ground, but they already used most of it, and cost and time for additional fiber installation prohibited expansion in many cases. The capacity of commercial optical fiber communications systems has increased at the speed of Moore's Law for integrated circuits, doubling every two years, as shown in Figure 16.2. The recent rate of increase for experimental systems accelerates

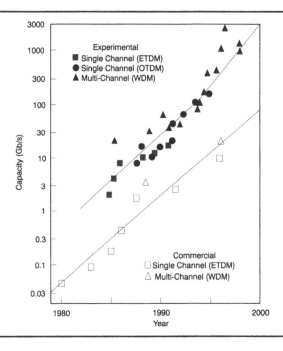

Figure 16.2: Capacity of Optical Fiber Over Time

HDTV and Digital Video

Industries such as the broadcast industry have begun using fiber optics. Today, copper cable can easily send studio-quality analog video signals a few hundred feet, making fiber unable compete with the cost of a 10-foot piece of copper coax cable. Sure, fiber optics offers better signal quality, lower weight, EMI immunity and so on, but if copper does the job at 1/10th the cost of fiber optics, installers will choose copper.

Moving to HDTV and digitized video, which require substantially more bandwidth, represent the changes facing the broadcast industry. Either of these changes could alter the copper/fiber cost balance. Curiously, long copper coax cable complicates digitized video more so than analog video. Embracing these new technologies increases fiber's acceptance in the broadcast industry. While fiber optics will play a big part in the broadcast studio, especially in studio-to-transmitter links, illustrated in Figure 16.3, and studio-to-studio links. However, its role in mass video distribution to the home remains uncertain. These technologies, even copper cable, require digital compression of video signals to effectively compete for this market.

Figure 16.3: Studio-to-Transmitter Link (STL) Application

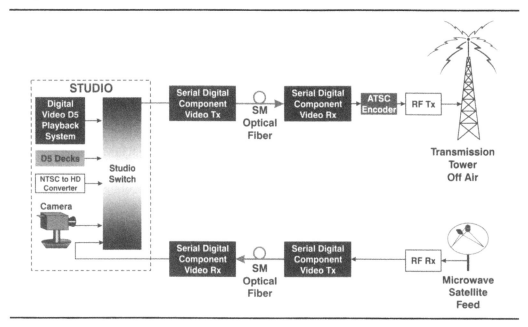

Smart Highways

Fiber optics improve transportation and highway monitoring. Smart highway systems often employ many sensors, communications stations, and cameras, all ideal applications for fiber. Increased public safety concern has prompted many cities to install hundreds of cameras in and around the city, allowing the police to constantly monitor these areas. The distances involved and immunity to lightning strikes make fiber a logical choice.

The prediction that fiber optics would penetrate process control applications has yet to be realized. Why? Again, the need and cost factors drove the choice to copper cable. On the other hand, new markets such as distance learning appear well-suited for fiber optics.

Figure 16.4: Variable Message Sign Typical in Smart Highway Systems

(Photo by Force, Inc.)

Many markets for fiber optics may not yet exist, making it impossible to predict the key future mass markets. Many once exotic applications have become commonplace in the last five years such as the information superhighway, smart highways, and distance learning. The same will happen in the next five years.

State-of-the-Art Telecommunications Networks

In the early 1990's, SONET systems using OC-1 (51.84 Mb/s) carried much of the long-haul telecommunications traffic. Today, the highest standards under consideration include OC-768 (39.808 Gb/s). In the mid 1990's the fastest data rate used commercially transmitted up to 2.5 Gb/s, also referred to as OC-48. Valid SONET data rates increase by a factor of four starting at OC-3 and increasing to OC-768. The next logical increase, 10 Gb/s or OC-192, became available in 1997.

The industry found it highly desirable to put more than one channel of information onto a single fiber. This reduced or eliminated the need to deploy more fiber. In addition, the maximum distance of a fiber optic transmission system (FOTS) needed to increase dramatically. This would reduce the overall cost of the data transmission link by eliminating or reducing the need for costly signal regeneration facilities along the length of the transmission path. The gap between the unfilled growing needs of the telecommunications industry and the growing demands for more bandwidth created a huge opportunity for fiber optics to make a breakthrough in performance and cost.

Several pieces of new fiber optic technology including DWDM, EDFA's, solitons, dispersion compensation, FEC coding and many more, left the hands of researchers and rapidly converted into next-generation FOTS. The next generation provides more than 100 times the capacity on one fiber compared to existing hardware, and at the same time, substantially increases the maximum allowable transmission distances. In many ways, the telecommunications industry adopted much of the technology deployed in submarine fiber optic systems for years. Submarine fiber optic systems, which must reliably run for thousands of miles underwater, dominate the cutting edge of technology of fiber optic transmission systems.

Terrestrial telecommunications industries ignored the more exotic technologies associated with submarine systems until the Internet created a need to dramatically increase the capacity of terrestrial systems. In the late 1990's, many large telecommunications equipment suppliers recognized the looming financial opportunity for greatly expanded capacity and distance on the fibers that they already deployed. Many companies have proposed new systems to carry a large number (over 100) of high data rate (10 Gb/s) telecommunication signals, over very large distances (more than 1,000 km), using DWDM, on one "standard" single-mode fiber. These future systems require a number of state-of-the-art technology building blocks to assemble a single transmission platform. DWDM systems in 1998 did not commercially exploit their ability to use the "standard" non dispersion-shifted fiber that had already been installed, but future trends in telecommunications networks will probably take advantage of this fiber, finding ways to make it transmit more channels.

COST REDUCTION

As mentioned earlier, the chicken and egg syndrome has inhibited fiber's mass deployment. (No doubt the same could be said for most new technologies.) In spite of the difficulties, all types of fiber optic components and systems have seen dramatic decreases in cost during the last decade, and there seems no leveling of this trend. The cost of key fiber optic components such as lasers, LED's and detectors has dropped at a yearly rate of 5-20%. Fiber, often a fraction of the system cost, has leveled off due to a shortage of capacity. Fiber optic connectors continue to make big strides in cost reduction due to innovative, simplified designs, and the introduction of new materials. Cost reduction allows fiber to move into expanded markets.

MINIATURIZATION

Making smaller components emerges as a common theme in many fiber optic applications. While some products reach the limit of diminishing returns now, others will continue. The size of fiber optic connectors limits the ability to miniaturize the component. Recent revivals of ribbon cable and mass termination connectors may answer that issue. The availability of highly integrated electronic circuits will also allow for continued advances in miniaturization. For example, the data link on the left in Figure 16.5 weighs ten pounds and transmits data at a rate of 20 Mb/s, a cutting edge data rate twenty years ago. By contrast, the link on the right, developed in late 1990's, transmits 1.5 Gb/s, weighs only 3 ounces, and measures less than two inches long.

Figure 16.5:
Miniaturization in Data Links

(Photos by Force, Inc.).

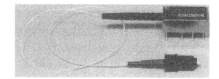

While miniaturization will continue, the point of diminishing returns has nearly been reached. The diversity of the fiber optics industry would make the development of a "complete link on a chip" unaffordable because the sales volumes may not be sufficient to pay the development costs. Because of this, the level of integration may stay at today's building block level for all but the highest-volume products.

NEW MATERIALS

In the early days of fiber optics, components tended to use expensive and exotic materials. Fiber optic connectors initially used stainless steel ferrules or even precision drilled jewels to achieve alignment. Later, most connectors changed to ceramics. Today, fiber optic connectors with plastic ferrules, unthinkable only a decade ago, are becoming more prevalent. Other fiber optic components evolved from complex laser-welded, hermetically sealed enclosures to low-cost, unsealed enclosures. Improvements in the environmental tolerance of the basic components enabled them to brave the elements with minimum packaging protection.

Trends like the widespread use of plastics in fiber optic components and connectors will continue. Epoxies, once frowned upon in many fiber optic assemblies, make acceptable alternatives to laser welding in some applications.

TECHNOLOGY REFINEMENTS

When a technology reaches the level of maturity that the fiber optics industry has reached, improvements in the technology tend to be evolutionary rather than revolutionary. This means that small incremental changes and improvements will be made to products each generation, much like the automobile industry.

Laser diodes represent a technology that moved from its revolutionary stage to its evolutionary stage look at what has happened with laser diodes in the fiber optics industry. Over the years dozens of breakthrough laser designs have come and gone. Fabry-Perot designs, cleaved-coupled designs, DFB designs exemplify the many designs along the way. Some designs died from starvation in the open market because of poor design quality. Some have thrived and formed the basis for improved designs.

All things considered, this evolutionary mode is good for the fiber optics industry as well as its customers. Things will not change so fast that manufacturers cannot recoup development costs or customers cannot keep up with the technology. Over 90% of the fiber optic systems that will be deployed over the next decade will be based on technology that is already in manufacturing today. Much of the real innovation will be at the system level, not the underlying component technology. Look at the DVD player. Lasers have existed for years as have techniques for digitizing video and audio, but putting all of these elements together to create the DVD player was revolutionary. The same will happen in the fiber optics industry.

NEW TECHNOLOGY

Here's the promise that excites everyone. The new glimmering bits of technology get the press, and investors, and potential customers inspired. In reality, only 10% or less of the "breakthroughs" ever get out of the lab. It's hard to know which to bet on, but the marketplace usually decides quickly. If a breakthrough doesn't yield better performance *and* lower cost, it will die. And while you tally up the score of a new piece of technology, don't forget to consider the competing technologies ever present and ever improving. Fiber optic sensors provide great example of this.

In the early 1980's fiber optic sensors were touted as the solution to the world's measurement problems. Here finally, was a lightweight, small, EMI resistant sensor. However, most of the people developing such sensors had no idea what existing technologies could already do. While the fiber optic sensor industry struggled to make a pressure sensor that could achieve 1% accuracy over a 0°C to +40°C temperature range for less than $1,000, you could buy a state-of the-art capacitive pressure sensor for $200 that guaranteed 0.01% accuracy from -20°C to +70°C. Despite this, an interest in fiber optic sensors remains in specialty applications.

Hope continues for long wavelength optics. Theory states that if you could operate a fiber at a wavelength of 3 μm to 4 μm, rather than the 1.3 μm or 1.55 μm wavelengths widely used today, fiber losses could drop to 0.001 dB/km. That means that a single continuous fiber stretched around the equator would have a loss of only 40 dB, achievable without the use of repeaters. However, serious practical problems abound. Fluorine based glass used in such fibers reacts destructively with water, making connections challenging. After a decade of trying, researchers have failed to get close to the theoretical loss of 0.001 dB/km raising serious doubts about the feasibility. Also the lasers, LED's or detectors would likely need to operate at liquid nitrogen temperatures.

Coherent communications has also been publicized as the future of fibers optics. Figure 16.6 illustrates this transmission scheme. In this system, the laser transmitter emits light at one frequency which is modulated using amplitude, phase or frequency modulation, by the signal. At the receive end, the light mixes with light from a second laser, and the intermediate frequency is then detected and converted to an output signal. This scheme avoids noise encountered in direct detection, making the receiver sensitive to faint signals which in turn increases the allowable optical loss between the transmitter and the receiver. This greater sensitivity in coherent communications should allow much higher data rates than the non-coherent communications used today. Increased sensitivity means the receiver requires fewer photons of light to detect the optical signal. One problem to this transmission scheme is that today's electronics cannot run fast enough to take full advantage of the technique. In the meantime, don't write off non-coherent communications. It will continue to evolve year after year taking incremental advantage of the improvements in electronics components and other pieces of technology. As a result, the edge offered by coherent communications will decline.

Figure 16.6:
Coherent
Transmission Scheme

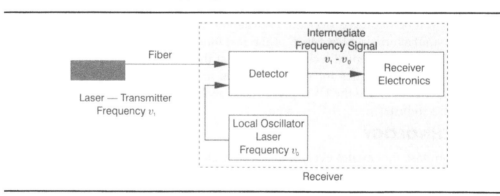

Integrated optics, which can broadly refer to two technologies, will certainly play a part in fiber's future. Integrating light emitters and/or detectors onto a substrate that holds some or all of the signal processing circuitry represents the first technology. However, existing components achieve a higher performance. Yield problems also drive up development costs. Perhaps in the future the technology can overcome these obstacles, but this technology must jump ahead of the declining price curve followed by other technologies such as discrete optic and electronic designs.

The creation of optical splitters and WDM's in single substrates represents the second technology. This differs significantly from the bulk optical techniques used in most existing devices. Again plagued by performance and yield problems, as manufacturers overcome production difficulties, these IOC's power the high-speed networks that make up the information superhighway.

The future of fiber optics will certainly be exciting. How closely it will resemble the viewpoints given here remains to be seen. Refinements in existing technology will most likely have the greatest impact, but new technology, still in the labs today, may yet emerge as new real-world applications. Those that do will certainly change the industry but only if they offer better performance and lower cost than existing and competing technologies. As is the case with most things, time will tell.

Chapter Summary

- Major trends for the future of fiber optics include increased fiber transparency, all-optical networks, multi-terabit transmission, competing technologies, expansion into mass markets, cost reduction, miniaturization, new materials, technology refinements and new technology.

- Exploiting a fifth window in optical fiber, which would stretch from about 1.35 μm to 1.53 μm, would give a continuous wavelength band that provides nearly 50 THz of bandwidth.

- All-optical networks do not require the data signal to be converted to the electrical domain in order to be switched or processed.

- Multi-terabit transmission would push data rates from the current 1 Tb/s up to 5 Tb/s or higher.

- The improvement of competing technologies such as digital video transmission over copper cable will reduce fiber's opportunities.

- Market expansion for fiber optics includes applications with the Internet, HDTV, smart highways, and state-of-the-art telecommunications networks.

- Technology refinements for the fiber optics industry have reached the level of maturity where they are evolutionary rather than revolutionary.

- New technology being developed in today's test labs will only be embraced if it offers better performance and lower cost than today's technology.

Selected References and Additional Reading

Brinkman, W. F and D. V. Lang. "Physics and the Communications Industry." Reviews of Modern Physics. 1999.

Emerging Technologies Research Group. "Internet Timeline." http://www.etrg.findsvp.com/time-line. 1998.

Gromov, Gregory R. "The Roads and Crossroads of Internet History." http://www.internetvalley.com. 1998.

Higgins, Thomas V. "Optoelectronics: The Next Technological Revolution." *Laser Focus World.* November 1995: 93-102.

Kopakowski, Edward T. "Metropolitan Area Networks for Interactive Distance Learning." *Fiberoptic Product News.* September 1995: 17+.

Lewis, Kirk. "Integrated Optics Building Better Products." *Photonics Spectra.* June 1995: 119-124.

Newell, Wade S. "Multi-Protocol High-Performance Serial Digital Fiber Optic Data Links," from *Digital Communications Design Conference, Day 3,* Joseph F. Havel, ed. Proceedings of Design SuperCon '95, Santa Clara, CA, Feb. 28-Mar. 2, 1995.

Refi, James J. "Optical Fibers for Optical Networking." *Bell Labs Journal.* Jan.-Mar., 1999: 246-261.

Steinberg, Mathew and John Ryan. "The High-Capacity Explosion." from *NFOEC Technical Proceedings, Volume 1.* Proceedings of the National Fiber Optic Engineers Conference, Orlando, FL, Sept. 13-17, 1998.

Yeh, Chai. *Handbook of Fiber Optics: Theory and Applications.* New York: Academic Press, Inc., 1990.

APPENDICES

GENERAL REFERENCE MATERIAL

CONCERNING NUMBERS

One encounters an extreme range of numbers in the fiber optics business. For example, a high-speed data link may respond in the picosecond range while a low-speed RS-232 link may respond in the millisecond range. To understand fiber optics and much of electronics, one must be familiar with the prefixes used to express these numbers. For example, deep red light at one end of the visible spectrum has a wavelength of 700 nanometers (nm) which is 700 billionths of a meter or 0.00002756 inches. Deep violet visible light has a wavelength roughly half of red's: 350 nm or 0.00001378 inches. Wavelengths of light are given in small numbers. If, however, we express light as a frequency, we get a large number. Deep red has a frequency of 430 THz or 430,000,000,000,000 Hz. Violet's frequency is 850 THz or 850,000,000,000,000 Hz.

Système Internationale or SI Units, more commonly referred to as the metric system, use prefixes to express large and small numbers. The following prefixes are used most.

Prefix	Symbol	Multiplier	Scientific	U.S. Word
exa	E	1,000,000,000,000,000,000	10^{18}	quintillions
peta	P	1,000,000,000,000,000	10^{15}	quadrillions
tera	T	1,000,000,000,000	10^{12}	trillions
giga	G	1,000,000,000	10^{9}	billions
mega	M	1,000,000	10^{6}	millions
kilo	k	1,000	10^{3}	thousands
hecto	h	100	10^{2}	hundreds
deka	da	10	10^{1}	tens
deci	d	0.1	10^{-1}	tenths
centi	c	0.01	10^{-2}	hundredths
milli	m	0.001	10^{-3}	thousandths
micro	μ	0.000 001	10^{-6}	millionths
nano	n	0.000 000 001	10^{-9}	billionths
pico	p	0.000 000 000 001	10^{-12}	trillionths
femto	f	0.000 000 000 000 001	10^{-15}	quadrillionths
atto	a	0.000 000 000 000 000 001	10^{-18}	quintillionths

Table A.1: Names and Symbols for Metric Prefixes

Mil	1/1000 of an inch	1 mil = 25.4 microns
Micron	1/1,000,000 of a meter	1 micron = 1 micrometer (μm)

CONSTANTS

The following is a partial list of constants that are pertinent to the use of fiber optics and electronics:

1. Speed of Light in a Vacuum
   C = 299,792.5 km/s = 29.98 cm/ns = 299.8 μm/ps

2. Speed of Light in Fiber (Refractive index = 1.45)
   C = 2.068 x 10^{8} m/s = 20.68 cm/ns = 206.8 μm/ps = 0.6783 ft./ns

3. Electronic Charge
   q = 1.602 x 10^{-19} Coulombs

4. **Planck's Constant**

   $h = 6.625 \times 10^{-34}$ J•s $= 4.135 \times 10^{-15}$ eV•s $= 6.625 \times 10^{-27}$ erg•s

5. **Boltzman's Constant**

   $k = 1.3804 \times 10^{-23}$ J/°K $= 8.616 \times 10^{-5}$ eV/°K $= 1.3804 \times 10^{-16}$ erg/°K

6. **Energy-Wavelength Conversion Factor**

   $hc/e = 1239.8$ eV nm $= 1.2398$ eV•μm

7. **Wien's Displacement Law Constant**

   λ_{max}•T $= 2.9878 \times 10^6$ nm•°K $= 2897.8$ μm•°K

8. **kT at +300°K**

   0.0259

9. **Characteristic Impedance of Vacuum**

   $Z_0 = 376.7$ Ω

10. **Pi (π)**

    3.1415926535

CONVERSION FACTORS

The following is a list of common conversions that pertain to fiber optics and electronics:

Table A.2: Conversion Factors

To Convert From:	To:	Multiply by:
μm (micrometers)	mils	0.03937
	inches	0.00003937
	Angstroms	10,000
	nanometers	1,000
mm (millimeters)	inches	0.03937
	feet	0.003281
cm (centimeters)	inches	.39370
	feet	0.03281
m (meters)	feet	3.281
	inches	39.37
	cm	100
	mm	1,000
km (kilometers)	miles	0.62137
	feet	3,280.8
	mm	1,000,000
kg (kilograms)	pounds	2.2046
g (grams)	oz. (ounces)	0.03527
	pounds	0.0022045
°C (Centigrade)	°F (Fahrenheit)	1.8 then add 32
°F	°C	subtract 32 then divide by 1.8
°K (Kelvin)	°C	= °C + 273.16
N (Newton)	pounds	0.2247
kg/km	pounds/mile	3.5480
kPa	PSI	0.14511

GLOSSARY OF TERMS

A Abbreviation for Ampere or Amp. The steady current produced by one Volt across a resistance of one Ohm.

Absorption That portion of optical attenuation in optical fiber resulting from the conversion of optical power to heat, caused by impurities in the fiber such as hydroxyl ions.

AC Abbreviation for alternating current. An electric current that reverses its direction at regularly recurring intervals

Acceptance Angle The half-angle of the cone within which incident light is totally internally reflected by the fiber core. It is equal to $\sin^{-1}$(NA).

Active Device A device that requires a source of energy for its operation and has an output that is a function of present and past input signals. Examples include controlled power supplies, transistors, LED's, amplifiers, and transmitters.

ACTS Abbreviation for advanced communications technologies and services. A program for developing long-haul telecommunications networks.

A/D or ADC Abbreviation for analog-to-digital converter. A device used to convert analog signals to digital signals.

Add-Drop Multiplexer (ADM) A device which adds or drops signals from a communications network.

ADSL Abbreviation for asymmetric digital subscriber line. See DSL.

Aerial Plant Cable that is suspended in the air on telephone or electric utility poles.

AGC Automatic gain control. A process or means by which gain is automatically adjusted in a specified manner as a function of input level or another specified parameter.

AM (Amplitude Modulation) A transmission technique in which the amplitude of the carrier is varied in accordance with the signal.

Amplified Spontaneous Emission (ASE) A background noise mechanism common to all types of erbium-doped fiber amplifiers (EDFA's). It contributes to the noise figure of the EDFA which causes loss of signal-to-noise ratio (SNR).

Amplifier A device that boosts the strength of an electronic signal. In a fiber optic cable system, amplifiers spaced at regular intervals throughout the system maintain signal fidelity.

Analog A continuously variable signal. A mercury thermometer, which gives a variable range of temperature readings, is an example of an analog instrument. Opposite of digital.

Angstrom (Å) A unit of length in optical measurements where:

$$1\text{Å} = 10^{-10} \text{ meters}$$
$$\text{or} = 10^{-4} \text{ micrometers}$$
$$\text{or} = 10^{-1} \text{ nanometers}$$

The angstrom has been used historically in the field of optics, but it is not an SI (Système Internationale or International System) unit. Rarely used in fiber optics, nanometers is preferred.

Angular Misalignment Loss at a connector due to fiber end face angles being misaligned.

ANSI Abbreviation for American National Standards Institute.

APC Abbreviation for angle polished connector. The fiber end face is polished at a 5°-15° angle for the minimum possible backreflection.

APL Abbreviation for average picture level. A video quality parameter.

APON Abbreviation for ATM passive optical network. A passive optical network operating in asynchronous transfer mode (ATM).

AR Coating Antireflection coating. A thin, dielectric or metallic film applied to an optical surface to reduce its reflectance and thereby increase its transmittance.

Armor A protective layer, usually metal, wrapped around a cable.

ASCII American standard code for information interchange. An encoding scheme used to interface between data processing systems, data communication systems, and associated equipment.

ASE See amplified spontaneous emission.

ASIC Abbreviation for application-specific integrated circuit. A custom-designed integrated circuit.

ASTM Abbreviation for American Society for Testing and Materials.

Asynchronous Data that is transmitted without an associated clock signal.

ATM (Asynchronous Transfer Mode) A transmission standard widely used by the telecom industry. A digital transmission switching format with cells containing 5 bytes of header information followed by 48 data bytes. Part of the B-ISDN standard.

ATE Abbreviation for automatic test equipment. Test equipment computer programmed to perform a number of test measurements on a device without the need for changing the test setup.

Attenuation The decrease in signal strength along a fiber optic waveguide caused by absorption and scattering. Attenuation is usually expressed in dB/km.

Attenuation Constant For a particular propagation mode in an optical fiber, the real part of the axial propagation constant.

Attenuation-limited Operation The condition in a fiber optic link when operation is limited by the power of the received signal (rather than by bandwidth or distortion).

Attenuator In optical or electrical systems, a usually passive device or network for reducing the amplitude of a signal without distorting the waveform.

Avalanche Photodiode (APD) A photodiode that exhibits internal amplification of photocurrent through avalanche multiplication of carriers in the junction region.

Average Power The average level of power in a signal that varies with time.

AWG (Arrayed Waveguide Grating) A device, built with silicon planar lightwave circuits, that allows multiple wavelengths to be combined and separated in DWDM systems.

Axial Propagation Constant For an optical fiber, the propagation constant evaluated along the axis of a fiber in the direction of transmission.

Axis The center of an optical fiber.

B See bel.

BB-I Abbreviation for broadband interactive services. The delivery of all types of interactive video, data, and voice services over a broadband communications network.

Backreflection In cases where light is launched into an optical fiber, backreflection refers to the light that is returned to the launch point.

Backscattering The return of a portion of scattered light to the input end of a fiber; the scattering of light in the direction opposite to its original propagation.

Bandgap Energy Energy gap in semiconductors and insulators. The forbidden energy level separating the valence band and the conduction band in a semiconductor or insulator. Expressed in electron Volts (eV).

Bandwidth The range of frequencies within which a fiber optic waveguide or terminal device can transmit data or information.

Bandwidth-limited Operation The condition in a fiber optic link when bandwidth, rather than received optical power, limits performance. This condition is reached when the signal becomes distorted, principally by dispersion, beyond specified limits.

Baseband A method of communication in which a signal is transmitted at its original frequency without being impressed on a carrier.

Baud A unit of signaling speed equal to the number of signal symbols per second, which may or may not be equal to the data rate in bits per second.

Beamsplitter An optical device, such as a partially reflecting mirror, that splits a beam of light into two or more beams. Used in fiber optics for directional couplers.

Bel (B) The logarithm to the base 10 of a power ratio, expressed as $B = \log_{10}(P_1/P_2)$, where P_1 and P_2 are distinct powers. The decibel, equal to one-tenth bel, is a more commonly used unit.

Bending Loss Attenuation caused by high-order modes radiating from the outside of a fiber optic waveguide as a result of a bend in the fiber. *See also* macrobending, microbending.

Bend Radius The smallest radius an optical fiber or fiber cable can bend before increased attenuation or breakage occurs.

BER (Bit Error Rate) The fraction of bits transmitted that are received incorrectly.

BIDI Abbreviation for bidirectional transceiver, a device that sends information in one direction and receives information from the opposite direction.

Bidirectional Operating in both directions.

Birefringent Having a refractive index that differs with respect to different polarizations of light.

Binary Base two numbers with only two possible values, 0 or 1. Primarily used by communication and computer systems.

Bit The smallest unit of information carried by a electrical or optical pulse, the basis for digital communications. Also refers to the pulse carrying the information.

BITE Built-in test equipment. Features designed into a piece of equipment that allow on-line diagnosis of failures and operating status.

BNC Popular coax bayonet style connector. Often used for baseband video.

BR See backreflection.

Bragg Grating A technique for building optical filtering functions directly into a piece of optical fiber based on interferometric techniques. Exposure to deep UV light through a grating makes the fiber photosensitive, forming regions of higher and lower refractive indices in the fiber core.

Broadband A method of communication where the signal is transmitted by being impressed on a high-frequency carrier.

Buffer *1.* A protective coating applied directly to optical fiber. 2. A computer routine or memory storage area used to compensate for a difference in rate of flow of data, or time of occurrence of events, when transferring data from one device to another.

Bus Network A network topology in which all terminals are attached to a transmission medium serving as a bus.

Butt Splice Two fibers joined without optical connectors and arranged end-to-end by means of a coupling. Fusion splicing is an example.

BW See bandwidth.

Bypass The ability of a station to isolate itself optically from a network while maintaining the continuity of the cable plant.

Byte A unit of eight bits.

C Abbreviation for Celsius. Measure of temperature where pure water freezes at 0° and boils at 100°.

Cable One or more optical fibers enclosed, with strength members, in a protective coverings.

Cable Assembly A cable that is connector terminated and ready for installation.

Cable Plant All of the optical elements, including fiber connectors, splices, etc., between a transmitter and a receiver.

Cable Television A communications system that distributes broadcast signals, non-broadcast signals, multiple satellite signals, original programming, and other signals by means of a coaxial cable and/or optical fiber.

Carrier-to-Noise Ratio The ratio, in decibels, of the level of the carrier to that of the noise in a receiver's IF bandwidth before any nonlinear process such as amplitude limiting and detection takes place.

CATV Originally an abbreviation for community antenna television, the term now typically refers to cable television. See cable television.

C-Band The wavelength range between 1530 nm and 1562 nm used in some CWDM and DWDM applications.

CCIR Abbreviation for Consultative Committee on Radio.

CCITT Abbreviation for Consultative Committee on Telephony and Telegraphy.

CCTV Abbreviation for closed-circuit television. An arrangement in which programs are directly transmitted to specific users and not broadcast to the general public.

CD Abbreviation for compact disk. Often used to describe high-quality audio, CD-quality audio, or short-wavelength lasers; CD laser.

CDMA Abbreviation for code-division multiple access, A coding scheme in which multiple channels are independently coded for transmission over a single wideband channel using an individual modulation scheme for each channel.

Center Wavelength In a laser, the nominal value central operating wavelength defined by a peak mode measurement where the effective optical power resides. In an LED, the average of the two wavelengths measured at the half amplitude points of the power spectrum.

Central Office A common carrier switching office in which users' lines terminate. The nerve center of a communications system.

CGA Abbreviation for color graphics adapter. A low-resolution standard for computer monitors.

Channel A communications path or the signal sent over that path. By multiplexing several channels, voice channels can be transmitted over an optical channel.

Channel Capacity Maximum number of channels that a cable system can carry simultaneously.

Chirp In laser diodes, the shift of the laser's central wavelength during single pulse durations due to laser instability.

Chromatic Dispersion Reduced optical signal strength caused by different wavelengths of light traveling at different speeds down the optical fiber. Chromatic dispersion occurs because the speed at which an optical pulse travels depends on its wavelength, a property inherent to all optical fiber. May be caused by material dispersion, waveguide dispersion, and profile dispersion.

Circulator Passive three-port devices that couple light from Port 1 to 2 and Port 2 to 3 and have high isolation in other directions.

Cladding Material that surrounds the core of an optical fiber. Its lower index of refraction, compared to that of the core, causes the transmitted light to travel down the core.

Cladding Mode A mode confined to the cladding; a light ray that propagates in the cladding.

Cleave The process of separating an optical fiber by a controlled fracture of the glass, for the purpose of obtaining a fiber end, which is flat, smooth, and perpendicular to the fiber axis.

cm Abbreviation for centimeter. Approximately 0.4 inches.

CMOS Complementary metal oxide semiconductor. A family of integrated circuits (IC's). Particularly useful for low to medium speed or low-power applications.

CMTS Abbreviation for cable modem termination system.

CNR See carrier-to-noise ratio.

CO See central office.

Coarse Wavelength-division Multiplexing (CWDM) CWDM allows eight or fewer channels to be stacked in the 1550 nm region of optical fiber, the C-Band.

Coating The material surrounding the cladding of a fiber. Generally a soft plastic material that protects the fiber from damage.

Coherent Communications A communication system where the output of a local laser oscillator mixes optically with a received signal, and the difference frequency is detected and amplified.

Color Subcarrier The 3.58 MHz signal which carries color information in a TV signal.

Composite Second Order (CSO) An important distortion measure of analog CATV systems. It is mainly caused by second order distortion in the transmission system.

Composite Sync A signal consisting of horizontal sync pulses, vertical sync pulses, and equalizing pulses only, with a no-signal reference level.

Composite Triple Beat (CTB) An important distortion measure of analog CATV systems. It is mainly caused by third-order distortion in the transmission system.

Composite Video A signal which consists of the luminance (black and white), chrominance (color), blanking pulses, sync pulses, and color burst.

Compression A process in which the dynamic range or data rate of a signal is reduced by controlling it as a function of the inverse relationship of its instantaneous value relative to a specified reference level. Compression is usually accomplished by separate devices called compressors and is used for many purposes such as: improving signal-to-noise ratios, preventing overload of succeeding elements of a system, or matching the dynamic ranges of two devices. Compression can introduce distortion, but it is usually not objectionable.

Concatenation The process of connecting pieces of fiber together.

Concentrator A multiport repeater.

Concentricity The measurement of how well-centered the core is within the cladding.

Connector A mechanical or optical device that provides a demountable connection between two fibers or a fiber and a source or detector.

Connector Plug A device used to terminate an optical conductor cable.

Connector Receptacle The fixed or stationary half of a connection that is mounted on a panel/bulkhead. Receptacles mate with plugs.

Connector Variation The maximum value in dB of the difference in insertion loss between mating optical connectors (e.g., with remating, temperature cycling, etc.). Also called optical connector variation.

Converter A device attached between the television set and the cable system that can increase the number of channels available on the TV set to accommodate the multiplicity of channels offered by cable TV.

Core The light-conducting central portion of an optical fiber, composed of material with a higher index of refraction than the cladding. The portion of the fiber that transmits light.

Counter-rotating An arrangement whereby two signal paths, one in each direction, exist in a ring topology.

Coupler An optical device that combines or splits power from optical fibers.

Coupling Ratio/Loss (C_R, C_L) The ratio/loss of optical power from one output port to the total output power, expressed as a percent. For a 1 x 2 WDM or coupler with output powers O_1 and O_2, and O_i representing both output powers:

Eq. B.1

$$C_R (\%) = \left(\frac{O_i}{O_1 + O_2}\right) \times 100\%$$

$$C_R (\%) = -10 \cdot \log_{10}\left(\frac{O_i}{O_1 + O_2}\right)$$

CPE See customer premises equipment.

Critical Angle In geometric optics, at a refractive boundary, the smallest angle of incidence at which total internal reflection occurs.

Cross-connect Connections between terminal blocks on the two sides of a distribution frame, or between terminals on a terminal block (also called straps). Also called cross-connection or jumper.

Cross-gain Modulation (XGM) A technique used in wavelength converters where gain saturation effects in an active optical device, such as a semiconductor optical amplifier (SOA), allow the conversion of the optical wavelength. Better at shorter wavelengths (e.g. 780 nm or 850 nm).

Cross-phase Modulation (XPM) A fiber nonlinearity caused by the index of refraction of glass. The index of refraction varies with optical power level causing different optical signals to interact.

Crosstalk (XT) 1. Undesired coupling from one circuit, part of a circuit, or channel to another. 2. Any phenomenon by which a signal transmitted on one circuit or channel of a transmission system creates and undesired effect in another circuit or channel.

CRT Abbreviation for cathode ray tube.

CSMA/CD Abbreviation for carrier sense multiple access with collision detection. A network control protocol in which (a) a carrier sensing is used and (b) a transmitting data station that detects another signal while transmitting a frame, stops transmitting that frame, waits for a jam signal, and then waits for a random time interval before trying to send that frame again.

CSO See composite second order.

CTB See composite triple beat.

CTS Abbreviation for clear to send. In a communications network, signal from a remote receiver to a transmitter that it is ready to receive a transmission.

Customer Premises Equipment (CPE) Terminal and associated equipment and inside wiring located at a subscriber's premises and connected with a carrier's communication channel(s) at the demarcation point ("demarc"), a point established in a building or complex to separate customer equipment from telephone company equipment. CPE does not include over-voltage protection equipment and pay telephones.

Cutback Method A technique of measuring optical fiber attenuation by measuring the optical power at two points at different distances from the test source.

Cutoff Wavelength In single-mode fiber, the wavelength below which the fiber ceases to be single-mode.

CW Abbreviation for continuous wave. Usually refers to the constant optical output from an optical source when it is biased (i.e., turned on) but not modulated with a signal.

CWDM See coarse wavelength-division multiplexing.

D/A or DAC Abbreviation for digital-to-analog converter. A device used to convert digital signals to analog signals.

Dark Current The induced current that exists in a reversed biased photodiode in the absence of incident optical power. It is better understood to be caused by the shunt resistance of the photodiode. A bias voltage across the diode (and the shunt resistance) causes current to flow in the absence of light.

Data Rate The number of bits of information in a transmission system, expressed in bits per second (b/s or bps), and which may or may not be equal to the signal or baud rate.

dB See decibel.

dBc Decibel relative to a carrier level.

dBμ Decibels relative to microwatt.

dBm Decibels relative to milliwatt.

DC Abbreviation for direct current. An electric current flowing in one direction only and substantially constant in value

DCE Abbreviation for data circuit-terminating equipment. *1.* In a data station, the equipment that (a) performs functions, such as signal conversion and coding, at the network end of the line between the data terminal equipment (DTE) and the line, and (b) may be a separate or an integral part of the DTE or of intermediate equipment. *2.* The interfacing equipment that may be required to couple the data terminal equipment (DTE) into a transmission circuit or channel and from a transmission circuit or channel into the DTE.

DCF See dispersion-compensating fiber.

DCM See dispersion-compensating module.

Decibel (dB) A unit of measurement indicating relative optic power on a logarithmic scale. Often expressed in reference to a fixed value, such as dBm (1 milliwatt) or dBμ (1 microwatt).

$$\text{Eq. B.2} \qquad\qquad dB = 10 \cdot \log_{10}\left(\frac{P_1}{P_2}\right)$$

Demultiplexer A module that separates two or more previously combined signals by compatible multiplexing equipment.

Dense Wavelength-Division Multiplexing (DWDM) This refers to the transmission of numerous closely spaced wavelengths in the 1550 nm region. Wavelength spacings are usually 100 GHz or 200 GHz which corresponds to 0.8 nm or 1.6 nm. DWDM bands include the C-Band, the S-Band, and the L-Band.

Detector An opto-electric transducer used in fiber optics to convert optical power to electrical current. Usually referred to as a photodiode.

DFB See distributed feedback laser.

DG See differential gain.

Diameter-mismatch Loss The loss of power at a joint, when coupling light from a source to a fiber, from fiber to fiber, or from fiber to detector, that occurs when the transmitting fiber has a diameter greater than the diameter of the receiving fiber.

Dichroic Filter An optical filter that transmits light according to wavelength. Dichroic filters reflect light that they do not transmit.

Dielectric Any substance in which an electric field may be maintained with zero or near-zero power dissipation. This term usually refers to non-metallic materials.

Differential Gain A type of distortion in a video signal that causes the brightness information to be distorted.

Differential Phase A type of distortion in a video signal that causes the color information to be distorted.

Diffraction Grating An array of fine, parallel, equally spaced reflecting or transmitting lines that mutually enhance the effects of diffraction to concentrate the diffracted light in a few directions determined by the spacing of the lines and by the wavelength of the light.

Digital A signal that consists of discrete states. A binary signal has only two states, 0 and 1.

Digital Compression An engineering technique for converting a cable television signal into a easily stored and manipulated digital format that requires a smaller portion of a spectrum for its transmission, carrying many channels in the capacity currently needed for one signal.

Diode An electronic device that lets current flow in only one direction. Semiconductor diodes used in fiber optics contain a junction between regions of different doping. They include light emitters (LED's and laser diodes) and detectors (photodiodes).

Diode Laser Synonymous with injection laser diode.

DIP Abbreviation for dual in-line package. An electronic package with a rectangular housing and a row of pins along each of two opposite sides.

Diplexer A device that combines two or more types of signals into a single output.

Directional Coupler A coupling device for separately sampling either the forward (incident) or the backward (reflected) wave in a transmission line through a known coupling loss.

Directivity See near-end crosstalk.

Dispersion The temporal spreading of a light signal in an optical waveguide caused by light signals traveling at different speeds through a fiber either due to modal or chromatic effects.

Dispersion-compensating Fiber (DCF) A fiber that has the opposite dispersion of the fiber being used in a transmission system. It is used to nullify the dispersion caused by that fiber.

Dispersion-compensating Module (DCM) This module has the opposite dispersion of the fiber being used in a transmission system. It is used to nullify the dispersion caused by that fiber. It can be either a spool of a special fiber or a grating based module.

Dispersion-shifted Fiber (DSF) A type of single-mode fiber designed to have zero dispersion near 1550 nm. This fiber type works very poorly for DWDM applications because of high fiber nonlinearity at the zero-dispersion point.

Dispersion Management A technique used in the system design of a fiber optic transmission to be able to cope with the dispersion introduced by the optical fiber.

Dispersion Penalty The loss of sensitivity in a receiver caused by dispersion in an optical fiber. Expressed in dB.

Distortion Nonlinearities in a unit that cause harmonics and beat products to be generated.

Distortion-limited Operation Generally synonymous with bandwidth-limited operation.

Distributed Feedback Laser (DFB) An injection laser diode which has a Bragg reflection grating in the active region in order to suppress multiple longitudinal modes and enhance a single longitudinal mode.

Distribution System Part of a cable system consisting of trunk and feeder cables used to carry signal from headend to customer terminals.

Dominant Mode The mode in an optical device spectrum with the most power.

Dope Thick liquid or paste used to prepare a surface or a varnish-like substance used for waterproofing or strengthening a material.

Double-Window Fiber An optical fiber optimized for use in two wavelengths. In multimode fibers, the fiber is optimized for 850 nm and 1310 nm operation. In single-mode fibers, the fiber is optimized for 1310 nm and 1550 nm operation.

DP See differential phase.

DSF See dispersion-shifted fiber.

DSL Abbreviation for digital subscriber line. In an integrated systems digital network (ISDN), equipment that provides full-duplex service on a single twisted metallic pair at a rate sufficient to support ISDN basic access and additional framing, timing recovery, and operational functions. *See also* integrated systems digital network

DSR Abbreviation for data signaling rate. The aggregate rate at which data pass a point in the transmission path of a data transmission system, expressed in bits per second (bps or b/s).

DST Abbreviation for dispersion supported transmission. In electrical TDM systems, a transmission system that would allow data rates at 40 Gb/s by incorporating devices such as SOA's.

DSx A transmission rate in the North American digital telephone hierarchy. Also called T-carrier.

DTE Abbreviation for data terminal equipment. *1.* An end instrument that converts user information into signals for transmission or reconverts the received signals into user information. *2.* The functional unit of a data station that serves as a data source or a data sink and provides for the data communication control function to be performed in accordance with link protocol.

DTR Abbreviation for data terminal ready. In a communications network, a signal from a remote transmitter that it is clear to receive data.

Dual Attachment Concentrator A concentrator that offers two attachments to the FDDI network which are capable of accommodating a dual (counter-rotating) ring.

Dual Attachment Station A station that offers two attachments to the FDDI network which are capable of accommodating a dual (counter- rotating) ring.

Dual Ring (FDDI Dual Ring) A pair of counter- rotating logical rings.

Duplex Cable A two-fiber cable suitable for duplex transmission.

Duplex Transmission Transmission in both directions, either one direction at a time (half-duplex) or both directions simultaneously (full-duplex).

Duty Cycle In a digital transmission, the fraction of time a signal is at the high level.

DWDM See dense wavelength-division multiplexing.

ECL Emitter-coupled logic. A high-speed logic family capable of GHz rates.

EDFA See erbium-doped fiber amplifier.

Edge-emitting Diode An LED that emits light from its edge, producing more directional output than surface-emitting LED's that emit from their top surface.

Effective Area The area of a single-mode fiber that carries the light.

EGA Abbreviation for enhanced graphics adapter. A medium-resolution color standard for computer monitors.

EIA Abbreviation for Electronic Industries Association.

8B10B Encoding A signal modulation scheme in which eight bits are encoded in a 10-bit word to ensure that too many consecutive zeroes do not occur; used in ESCON and fibre channel.

802.3 Network A 10 Mb/s CSMA/CD bus-based network; commonly called ethernet.

802.5 Network A token-passing ring network operating at 4 Mb/s or 16 Mb/s.

Electromagnetic Interference (EMI) Any electrical or electromagnetic interference that causes undesirable response, degradation, or failure in electronic equipment. Optical fibers neither emit nor receive EMI.

Electromagnetic Radiation (EMR) Radiation made up of oscillating electric and magnetic fields and propagated with the speed of light. Includes gamma radiation, X-rays, ultraviolet, visible, and infrared radiation, and radar and radio waves.

Electromagnetic Spectrum The range of frequencies of electromagnetic radiation from zero to infinity.

ELED See edge-emitting diode.

Ellipticity Describes a fiber's core or cladding as elliptical rather than circular.

EM Abbreviation for electromagnetic.

EMD See equilibrium mode distribution.

EMI Abbreviation for electromagnetic interference. *1.* Any electromagnetic disturbance that interrupts, obstructs, or otherwise degrades or limits the effective performance of electronics/electrical equipment. It can be induced intentionally, as in some forms of electronic warfare, or unintentionally, as a result of spurious emissions and responses, intermodulation products, and the like. *2.* An engineering term used to designate interference in a piece of electronic equipment caused by another piece of electronic or other equipment.

EMP Abbreviation for electromagnetic pulse. A burst of electromagnetic radiation that creates electric and magnetic fields that may couple with electrical/electronic systems to produce damaging current and voltage surges.

EMR Electromagnetic radiation. Radiation made up of oscillating electric and magnetic fields and propagated with the speed of light. Includes gamma radiation, X-rays, ultraviolet, visible, and infrared radiation, and radar and radio waves.

Endoscope A fiber optic bundle used for imaging and viewing inside the human body.

E/O Abbreviation for electrical-to-optical converter. A device that converts electrical signals to optical signals.

Equilibrium Mode Distribution (EMD) The steady modal state of a multimode fiber in which the relative power distribution among modes is independent of fiber length.

Erbium-doped Fiber Amplifier (EDFA) Optical fibers doped with the rare earth element erbium, which can amplify light in the 1550 nm region when pumped by an external light source.

ESCON Abbreviation for enterprise systems connection. A duplex optical connector used for computer-to-computer data exchange.

Ethernet A baseband local area network marketed by Xerox and developed jointly by Xerox, Digital Equipment Corporation, and Intel.

Evanescent Wave Light guided in the inner part of an optical fiber's cladding rather than in the core.

Excess Loss In a fiber optic coupler, the optical loss from that portion of light that does not emerge from the nominal operation ports of the device.

External Modulation Modulation of a light source by an external device that acts like an electronic shutter.

Extinction Ratio The ratio of the low, or OFF optical power level (P_L) to the high, or ON optical power level (P_H).

$$\text{Eq. B.3} \qquad \text{Extinction Ratio (\%)} = \left(\frac{P_L}{P_H}\right) \times 100$$

Extrinsic Loss In a fiber interconnection, that portion of loss not intrinsic to the fiber but related to imperfect joining of a connector or splice.

Eye Pattern Also called eye diagram. The proper function of a digital system can be quantitatively described by its BER, or qualitatively by its eye pattern. The "openness" of the eye relates to the BER that can be achieved.

F Abbreviation for Fahrenheit. Measure of temperature where pure water freezes at 32° and boils at 212°.

Failure Rate The number of failures of a device per unit of time.

Fall Time Also called turn-off time. The time required for the trailing edge of a pulse to fall from 90% to 10% of its amplitude; the time required for a component to produce such a result. Typically measured between the 80% and 20% points or alternately the 90% and 10% points.

FAR Abbreviation for Federal Acquisition Regulation. The guidelines by which the U.S. government purchases goods and services. Also the criteria that must be met by the vendor in order to considered as a source for goods and services purchased by the U.S. government.

Faraday Effect A phenomenon that causes some materials to rotate the polarization of light in the presence of a magnetic field parallel to the direction of propagation. Also called magneto-optic effect.

Far-end Crosstalk See wavelength isolation.

FBG Abbreviation for fiber Bragg gratings. See Bragg grating.

FC A threaded optical connector that originated in Japan. Good for single-mode or multimode fiber and applications requiring low backreflection.

FCC Abbreviation for Federal Communications Commission.

FC/PC See FC. A special curved polish on the connector for very low backreflection.

FCS Abbreviation for frame check sequence. An error-detection scheme that (a) uses parity bits generated by polynomial encoding of digital signals, (b) appends those parity bits to the digital signal, and (c) uses decoding algorithms that detect errors in the received digital signal.

FDA Abbreviation for Food and Drug Administration.

FDDI Abbreviation for fiber distributed data interface. *1.* A dual counter-rotating ring local area network. *2.* A connector used in a dual counter-rotating ring local area network

FDM See frequency-division multiplexing.

FEC See forward error correcting.

Ferrule A rigid tube that confines or holds a fiber as part of a connector assembly.

FET Abbreviation for field-effect transistor. A semiconductor so named because a weak electrical signal coming in through one electrode creates an electrical field through the rest of the transistor. This field flips from positive to negative when the incoming signal does, and controls a second current traveling through the rest of the transistor. The field modulates the second current to mimic the first one -- but it can be substantially larger.

Fiber Grating An optical fiber in which the refractive index of the core varies periodically along its length, scattering light in a way similar to a diffraction grating, and transmitting or reflecting certain wavelengths selectively.

Fiber Optic Attenuator A component installed in a fiber optic transmission system that reduces the power in the optical signal. It is often used to limit the optical power received by the photodetector to within the limits of the optical receiver.

Fiber Optic Cable A cable containing one or more optical fibers.

Fiber Optic Communication System The transfer of modulated or unmodulated optical energy through optical fiber media which terminates in the same or different media.

Fiber Optic Gyroscope A coil of optical fiber that can detect rotation about its axis.

Fiber Optic Link A transmitter, receiver, and cable assembly that can transmit information between two points.

Fiber Optic Span An optical fiber/cable terminated at both ends which may include devices that add, subtract, or attenuate optical signals.

Fiber Optic Subsystem A part of a system acting as a functional entity with defined bounds and interfaces. It contains solid state and/or other components and is specified as a subsystem for the purpose of trade and commerce.

Fiber-to-the-home (FTTH) Fiber optic service to a node located inside an individual home.

Fiber-to-the-loop (FTTL) Fiber optic service to a node that is located in a neighborhood.

Fibre Channel An industry-standard specification that originated in Great Britain which details computer channel communications over fiber optics at transmission speeds from 132 Mb/s to 1062.5 Mb/s at distances of up to 10 kilometers.

Filter A device which transmits only part of the incident energy and may thereby change the spectral distribution of energy.

FIT Rate Number of device failures in one billion device hours.

Fluoride Glasses Materials that have the amorphous structure of glass but are made of fluoride compounds (e.g., zirconium fluoride) rather than oxide compounds (e.g., silica). Suitable for very long wavelength transmission.

FM (Frequency Modulation) A method of transmission in which the carrier frequency varies in accordance with the signal.

Forward Error Correcting (FEC) A communication technique used to compensate for a noisy transmission channel. Extra information is sent along with the primary data payload to correct for errors that occur in transmission.

FOTP (Fiber Optic Test Procedure) Standards developed and published by the Electronic Industries Association (EIA) under the EIA-RS-455 series of standards. See Appendix D.

4B5B Encoding A signal modulation scheme used in FDDI that encodes groups of four bits transmits them in five bits in order to guarantee that no more than three consecutive zeroes ever occur.

Four Wave Mixing (FWM) A nonlinearity common in DWDM systems where multiple wavelengths mix together to form new wavelengths. Most prevalent near the zero-dispersion point and at close wavelength spacings.

FP Abbreviation for Fabry-Perot. Generally refers to any device such as a type of laser that uses mirrors in an internal cavity to produce multiple reflections.

Frequency-division Multiplexing (FDM) A method of deriving two or more simultaneous, continuous channels from a transmission medium by assigning separate portions of the available frequency spectrum to each of the individual channels.

Frequency-shift Keying (FSK) Frequency modulation in which the modulating signal shifts the output frequency between predetermined values. Also called frequency-shift modulation, frequency-shift signaling.

Fresnel Reflection Loss Reflection losses at the ends of fibers caused by differences in the refractive index between glass and air. The maximum reflection caused by a perpendicular air-glass interface is about 4% or about -14 dB.

FSAN Abbreviation for full service access network. A forum for the world's largest telecommunications services providers and equipment suppliers to work define broadband access networks based primarily on the ATM passive optical network structure.

FSK See frequency-shift keying.

FTTC See fiber-to-the-curb.

FTTH See fiber-to-the-home.

FTTL See fiber-to-the-loop.

Full-duplex Simultaneous bidirectional transfer of data.

Fused Coupler A method of making a multimode or single-mode coupler by wrapping fibers together, heating them, and pulling them to form a central unified mass so that light on any input fiber is coupled to all output fibers.

Fused Fiber A bundle of fibers fused together so that they maintain a fixed alignment with respect to each other in a rigid rod.

Fusion Splicer An instrument that permanently bonds two fibers together by heating and fusing them.

FUT Abbreviation for fiber under test. Refers to the fiber being measured by some type of test equipment.

FWHM Abbreviation for full width half maximum. Used to describe the width of a spectral emission at the 50% amplitude points. Also know as FWHP (full width half power).

FWM See Four Wave Mixing.

G Abbreviation for giga. One billion or 10^9.

GaAlAs Abbreviation for gallium aluminum arsenide. A semiconductor material used to make short wavelength light emitters.

GaAs Abbreviation for gallium arsenide. A semiconductor material used to make light emitters.

Gain The ratio of output current, voltage, or power to input current, voltage, or power, respectively. Usually expressed in dB.

GaInAsP Abbreviation for gallium indium arsenide phosphide. A semiconductor material used to make long wavelength light emitters.

Gap Loss Loss resulting from the end separation of two axially aligned fibers.

Gate 1. A device having one output channel and one or more input channels, such that the output channel state is completely determined by the input channel states, except during switching transients. 2. One of the many types of combinational logic elements having at least two inputs.

Gaussian Beam A beam pattern used to approximate the distribution of energy in a fiber core. It can also be used to describe emission patterns from surface-emitting LED's. Most people would recognize it as the bell curve. The equation that defines a Gaussian beam is:

$$\text{Eq. B.4} \qquad\qquad E(x) = E(0)e^{-x^2/W_0^2}$$

GBaud One billion bits of data per second or 10^9 bits.

Gb/s See GBaud.

Ge Abbreviation for germanium. An element used to make detectors. Good for most wavelengths (e.g., 800-1600 nm).

Genlock A process of sync generator locking. This is usually performed by introducing a composite video signal from a master source to the subject sync generator. The generator to be locked has circuits to isolate the subcarrier, vertical drive, and horizontal drive. The subject generator locks to the master subcarrier, horizontal, and vertical drives, resulting in both sync generators running at the same frequency and phase.

GHz Abbreviation for gigahertz. One billion Hertz (cycles per second) or 10^9 Hertz.

Graded-index Fiber Optical fiber in which the refractive index of the core is in the form of a parabolic curve, decreasing toward the cladding.

GRIN Gradient index. Generally refers to the SELFOC lens often used in fiber optics.

Ground Loop Noise Noise that results when equipment is grounded at points having different potentials, thereby creating an unintended current path. The dielectric properties of optical fiber provide electrical isolation that eliminates ground loops.

Group Index Also called group refractive index. In fiber optics, for a given mode propagating in a medium of refractive index (n), the group index (N), is the velocity of light in a vacuum (c), divided by the group velocity of the mode.

Group Velocity 1. The velocity of propagation of an envelope produced when an electromagnetic wave is modulated by, or mixed with, other waves of different frequencies. 2. For a particular mode, the reciprocal of the rate of change of the phase constant with respect to angular frequency. 3. The velocity of the modulated optical power.

Half-duplex A bidirectional link that is limited to one-way transfer of data, i.e., data can't be sent both ways at the same time.

Hard-clad Silica Fiber An optical fiber having a silica core and a hard polymeric plastic cladding intimately bounded to the core.

HBT Abbreviation for heterojunction bipolar transistors. A very high performance transistor structure built using more than one semiconductor material.

HDSL Abbreviation for high data-rate digital subscriber line. A DSL operating at a high data rate compared to the data rates specified for ISDN. *See also* DSL.

HDTV High-definition television. Television that has approximately twice the horizontal and twice the vertical emitted resolution specified by the NTSC standard.

Headend 1. A central control device required within some LAN/MAN systems to provide such centralized functions as remodulation, retiming, message accountability, contention control, diagnostic control, and access to a gateway. 2. A central control device within CATV systems to provide such centralized functions as remodulation. *See also* local area network (LAN).

HFC See hybrid fiber coax.

Hero Experiments Experiments performed in a laboratory environment to test the limits of a given technology.

Hertz One cycle per second.

HIPPI High performance parallel interface as defined by ANSI X3T9.3 document.

HP Abbreviation for homes passed.

Hybrid Fiber Coax (HFC) A cable construction that incorporates both fiber optic transmission components and copper coax transmission components.

Hydrogen Losses Increases in fiber attenuation that occur when hydrogen diffuses into the glass matrix and absorbs light.

Hz See Hertz.

IC See integrated circuit.

IDP See integrated detector/preamplifier.

IEEE Abbreviation for Institute of Electrical and Electronic Engineers.

IETF Abbreviation for Internet engineering task force.

IIN See interferometric intensity noise.

IM See intensity modulation.

Index-matching Fluid A fluid whose index of refraction nearly equals that of the fiber's core. Used to reduce Fresnel reflection at fiber ends. *See also* index-matching gel.

Index-matching Gel A gel whose index of refraction nearly equals that of the fiber's core. Used to reduce Fresnel reflection at fiber ends. *See also* index-matching fluid.

Index of Refraction Also refractive index. The ratio of the velocity of light in free space to the velocity of light in a fiber material. Symbolized by n. Always greater than or equal to one.

Infrared (IR) The region of the electromagnetic spectrum bounded by the long-wavelength extreme of the visible spectrum (about 0.7 μm) and the shortest microwaves (about 0.1 mm). *See also* frequency, light.

Infrared Fiber Colloquially, optical fibers with best transmission at wavelengths of 2 μm or longer, made of materials other than silica glass. *See also* fluoride glasses.

InGaAs Indium gallium arsenide. A semiconductor material used to make high-performance long-wavelength detectors.

InGaAsP Indium gallium arsenide phosphide. A semiconductor material used to make long-wavelength light emitters.

Injection Laser Diode (ILD) A laser employing a forward-biased semiconductor junction as the active medium. Stimulated emission of coherent light occurs at a pn junction where electrons and holes are driven into the junction.

In-line Amplifier And EDFA or other type of amplifier placed in the transmission line to amplify the attenuated signal before being sent onto the next, distant site. In-line amplifiers are all optical devices.

InP Indium phosphide. A semiconductor material used to make optical amplifiers and HBT's.

Insertion Loss The loss of power that results from inserting a component, such as a connector or splice, into a previously continuous path.

Integrated Circuit (IC) An electronic circuit that consists of many individual circuit elements, such as transistors, diodes, resistors, capacitors, inductors, and other active and passive semiconductor devices, formed on a single chip of semiconducting material and mounted on a single piece of substrate material.

Integrated Detector/Preamplifier (IDP) A detector package containing a PIN photodiode and transimpedance amplifier.

Integrated Services Digital Network (ISDN) An integrated digital network in which the same time-division switches and digital transmission paths are used to establish connections for services such as telephone, data, electronic mail and facsimile. How a connection is accomplished is often specified as a switched connection, nonswitched connection, exchange connection, ISDN connection, etc.

Intensity The square of the electric field strength of an electromagnetic wave. Intensity is proportional to irradiance and may get used in place of the term "irradiance" when only relative values are important.

Intensity Modulation (IM) In optical communications, a form of modulation in which the optical power output of a source is varied in accordance with some characteristic of the modulating signal.

Interchannel Isolation The ability to prevent undesired optical energy from appearing in one signal path as a result of coupling from another signal path. Also called crosstalk.

Interferometer An instrument that uses the principle of interference of electromagnetic waves for purposes of measurement. Used to measure a variety of physical variables, such as displacement (distance), temperature, pressure, and strain.

Interferometric Intensity Noise (IIN) Noise generated in optical fiber caused by the distributed backreflection that all fiber generates mainly due to Rayleigh scattering. OTDR's make use of this scattering power to deduce the fiber loss over distance.

Interferometric Sensors Fiber optic sensors that rely on interferometric detection.

Intermodulation (Mixing) A fiber nonlinearity mechanism caused by the power dependent refractive index of glass. Causes signals to beat together and generate interfering components at different frequencies.

International Telecommunications Union (ITU) A civil international organization, headquartered in Geneva, Switzerland, established to promote standardized telecommunications on a worldwide basis. The ITU-R and ITU-T are committees under the ITU which is recognized by the United Nations as the specialized agency for telecommunications.

Internet A worldwide collection of millions of computers that consists mainly of the WWW and e-mail.

Intrinsic Losses Splice losses arising from differences in the fibers being spliced.

IP Abbreviation for Internet protocol. A Department of Defense standard protocol for use in interconnected systems of packet-switched computer communications networks.

IPCEA Abbreviation for Insulated Power Cable Engineers Association.

IPI Intelligent peripheral interface as defined by ANSI X3T9.3 document.

IR See infrared.

Irradiance Power per unit area.

ISA Abbreviation for Instrument Society of America.

ISDN See integrated systems digital network.

ISP Abbreviation for Internet service provider. A company or organization that provides internet connections to individuals or companies via dial-up, ISDN, T1, or some other connection.

ISO Abbreviation for International Standards Organization.

Isolator A device that prevents backreflected light from re-entering an amplifier.

Isolation See near-end crosstalk.

ITU See International Telecommunications Union.

Jacket The outer, protective covering of a fiber optic or other type of cable.

Jitter Small and rapid variations in the timing of a waveform due to noise, changes in component characteristics, supply voltages, imperfect synchronizing circuits, etc.

Jitter, Data Dependent (DDJ) Also called data dependent distortion. Jitter related to the transmitted symbol sequence. DDJ is caused by the limited bandwidth characteristics, non-ideal individual pulse responses, and imperfections in the optical channel components.

Jitter, Duty Cycle Distortion (DCD) Distortion usually caused by propagation delay differences between low-to-high and high-to-low transitions. DCD is manifested as a pulse width distortion of the nominal baud time.

Jitter, Random (RJ) Random jitter is due to thermal noise and may be modeled as a Gaussian process. The peak-to-peak value of RJ is of a probabilistic nature, and thus any specific value requires an associated probability.

JPEG Abbreviation for Joint photographers expert group. International standard for compressing still photography.

Jumper A short fiber optic cable with connectors on both ends.

k Abbreviation for kilo. One thousand or 10^3.

K Abbreviation for Kelvin. Measure of temperature where pure water freezes at 273° and boils at 373°.

kBaud One thousand bits of data per second.

kb/s See kBaud.

Kevlar® A very strong, very light, synthetic compound developed by DuPont which is used to strengthen optical cables.

Keying Generating signals by the interruption or modulation of a steady signal or carrier.

kg Abbreviation for kilogram. Approximately 2.2 pounds.

kHz Abbreviation for kilohertz. One thousand cycles per second.

km Abbreviation for kilometer. 1 km = 3,280 feet or 0.62 miles.

Lambertian Emitter An emitter that radiates according to Lambert's cosine law. This law states that the radiance of certain idealized surfaces is dependent upon the angle from which the surface is viewed. The radiant intensity of such a surface is maximum normal to the surface and decreases in proportion to the cosine of the angle from the normal. Given by:

Eq. B.5
$$N = N_0 \cos(A)$$

Where:

N = The radiant intensity.

N_0 = The radiance normal (perpendicular) to an emitting surface.

A = The angle between the viewing direction and the normal to the surface.

LAN (Local Area Network) A communication link between two or more points within a small geographic area, such as between buildings.

Large Effective Area Fiber (LEAF) An optical fiber, developed by Corning, designed to have a large area in the core, which carries the light.

Large Core Fiber Usually, a fiber with a core of 200 μm or more.

Laser Acronym for light amplification by stimulated emission of radiation A light source that produces, through stimulated emission, coherent, near monochromatic light. Lasers in fiber optics are usually solid-state semiconductor types.

Laser Diode (LD) A semiconductor that emits coherent light when forward biased.

Lateral Displacement Loss The loss of power that results from lateral displacement of optimum alignment between two fibers or between a fiber and an active device.

Launch Fiber An optical fiber used to couple and condition light from an optical source into an optical fiber. Often the launch fiber is used to create an equilibrium mode distribution in multimode fiber. Also called launching fiber.

L-Band The wavelength range between 1570 nm and 1610 nm used in some CWDM and DWDM applications.

LD See laser diode.

LEAF See large effective area fiber.

LEC See local exchange carrier.

LED See light-emitting diode.

LEX Abbreviation for local exchange. *Synonym* central office.

LH Abbreviation for long-haul. A classification of video performance under RS-250C. Lower performance than medium-haul or short-haul.

L-I Curve The plot of optical output (L) as a function of current (I) which characterizes an electrical to optical converter.

Light In a strict sense, the region of the electromagnetic spectrum that can be perceived by human vision, designated the visible spectrum and nominally covering the wavelength range of 0.4 μm to 0.7 μm. In the laser and optical communication fields, custom and practice have extended usage of the term to include the much broader portion of the electromagnetic spectrum that can be handled by the basic optical techniques used for the visible spectrum. This region has not been clearly defined, but, as employed by most workers in the field, may be considered to extend from the near-ultraviolet region of approximately 0.3 μm, through the visible region, and into the mid-infrared region to 30 μm.

Light-emitting Diode (LED) A semiconductor that emits incoherent light when forward biased.

Light Piping Use of optical fibers to illuminate.

Lightguide *Synonym* optical fiber.

Light wave The path of a point on a wavefront. The direction of the light wave is generally normal to the wavefront.

Local Exchange Carrier (LEC) A local telephone company, i.e., a communications common carrier that provides ordinary local voice-grade telecommunications service under regulation within a specified service area.

Local Loop *Synonym* loop.

Long-haul Telecommunications Long-distance telecommunications links such as cross-country or transoceanic.

Longitudinal Mode An optical waveguide mode with boundary condition determined along the length of the optical cavity.

Loop *1.* A communication channel from a switching center or an individual message distribution point to the user terminal. *2.* In telephone systems, a pair of wires from a central office to a subscribers's telephone. *3.* Go and return conductors of an electric circuit; a closed circuit. *4.* A closed path under measurement in a resistance test. *5.* A type of antenna used extensively in direction-finding equipment and in UHF reception. *6.* A sequence of instructions that may be executed iteratively while a certain condition prevails.

Loose-tube A type of fiber optic cable construction where the fiber is contained within a loose tube in the cable jacket.

Loss The amount of a signal's power, expressed in dB, that is lost in connectors, splices, or fiber defects.

Loss Budget An accounting of overall attenuation in a system.

m Abbreviation for meter. 39.37".

M Abbreviation for mega. One million or 10^6.

mA Abbreviation for milliamp. One thousandth of an Amp or 10^{-3} Amps.

MAC *1.* Abbreviation for multiplexed analog components. A video standard developed by the European Community. An enhanced version, HD-MAC delivers 1250 lines at 50 frames per second, HDTV quality. *2.* Abbreviation for medium access control.

Macrobending In a fiber, all macroscopic deviations of the fiber's axis from a straight line.

MAN See metropolitan area network.

MAP Abbreviation for manufacturing automation protocol. Computer programs that run manufacturing automation systems.

Margin Allowance for attenuation in addition to that explicitly accounted for in system design.

Mass Splicing Simultaneous splicing of many fibers in a cable.

Material Dispersion Dispersion resulting from the different velocities of each wavelength in a material.

MBaud One million bits of information per second. Also referred to as Mbps or Mb/s.

Mb/s See MBaud.

Mean Launched Power The average power for a continuous valid symbol sequence coupled into a fiber.

Mechanical Splice An optical fiber splice accomplished by fixtures or materials, rather than by thermal fusion.

Medium Access Control 1. A service feature or technique used to permit or deny use of the components of a communication system. 2. A technique used to define or restrict the rights of individuals or application programs to obtain data from, or place data onto, a storage device, or the definition derived from that technique.

Metropolitan Area Network (MAN) A network covering an area larger than a local area network. A wide area network that covers a metropolitan area. Usually, an interconnection of two or more local area networks.

MFD See mode field diameter.

MH Abbreviation for medium-haul. A classification of video performance under RS-250C. Higher performance than long-haul and lower performance than short-haul.

MHz Abbreviation for megahertz. One million Hertz (cycles per second).

Microbending Mechanical stress on a fiber may introduce local discontinuities called microbending. This results in light leaking from the core to the cladding by a process called mode coupling.

Micrometer One millionth of a meter or 10^{-6} meters. Abbreviated μm.

Microsecond One millionth of a second or 10^{-6} seconds. Abbreviated μs.

Microwatt One millionth of a Watt or 10^{-6} Watts. Abbreviated μW.

MIL-SPEC Abbreviation for military specification.

MIL-STD Abbreviation for military standard.

Misalignment Loss The loss of power resulting from angular misalignment, lateral displacement, and end separation.

MLM See multilongitudinal mode laser.

mm Abbreviation for millimeter. One thousandth of a meter or 10^{-3} meters.

MM Abbreviation for multimode. See multimode.

Modal Dispersion See multimode dispersion.

Modal Noise Modal noise occurs whenever the optical power propagates through mode-selective devices. It is usually only a factor with laser light sources.

Mode A single electromagnetic wave traveling in a fiber.

Mode Coupling The transfer of energy between modes. In a fiber, mode coupling occurs until equilibrium mode distribution (EMD) is reached.

Mode Evolution The dynamic process a multilongitudinal laser undergoes whereby the changing distribution of power among the modes creates a continuously changing envelope of the laser's spectrum.

Mode Field Diameter (MFD) A measure of distribution of optical power intensity across the end face of a single-mode fiber.

Mode Filter A device that removes higher-order modes to simulate equilibrium mode distribution.

Mode Scrambler A device that mixes modes to uniform power distribution.

Mode Stripper A device that removes cladding modes.

Modulation The process by which the characteristic of one wave (the carrier) is modified by another wave (the signal). Examples include amplitude modulation (AM), frequency modulation (FM), and pulse-coded modulation (PCM).

Modulation Index In an intensity-based system, the modulation index is a measure of how much the modulation signal affects the light output. It is defined as follows:

$$\text{Eq. B.6} \qquad m = \frac{\text{highlevel-lowlevel}}{\text{highlevel+lowlevel}}$$

Monitor A television that receives its signal directly from a VCR, camera, or separate TV tuner for high-quality picture reproduction. Also a special type of television receiver designed for use with closed circuit TV equipment.

Monochrome Black and white TV signal.

MPEG Abbreviation for motion picture experts group. An international standard for compressing video that provides for high compression ratios. The standard has two recommendations: MPEG-1 compresses lower-resolution images for videoconferencing and lower-quality desktop video applications and transmits at around 1.5 Mb/s. MPEG-2 was devised primarily for delivering compressed television for home entertainment and is used at CCIR resolution when bit rates exceed 5.0 Mbits per second as in hard disk-based applications.

MQW See multi-quantum well laser.

ms Abbreviation for milliseconds. One thousandth of a second or 10^{-3} seconds.

MSO Abbreviation for multiple service operator. A telecommunication company that offers more than one service, e.g. telephone service, internet access, satellite service, etc.

MTBF Abbreviation for mean time between failure. Time at which 50% of the units of interest will have failed. Also called MTTF (mean time to failure).

Multilongitudinal Mode Laser (MLM) An injection laser diode which has a number of longitudinal modes.

Multimode Dispersion Dispersion resulting from the different transit lengths of different propagating modes in a multimode optical fiber. Also called modal dispersion.

Multimode Fiber An optical fiber that has a core large enough to propagate more than one mode of light. The typical diameter is 62.5 micrometers.

Multimode Laser Diode (MMLD) Synonym for multilongitudinal mode laser.

Multiple Reflection Noise (MRN) The fiber optic receiver noise resulting from the interference of delayed signals from two or more reflection points in a fiber optic span. Also known as multipath interference.

Multiplexer A device that combines two or more signals into a single output.

Multiplexing The process by which two or more signals are transmitted over a single communications channel. Examples include time-division multiplexing and wavelength-division multiplexing.

Multi-quantum Well Laser A laser structure with a very thin (about 10 nm thick) layer of bulk semiconductor material sandwiched between the two barrier regions of a higher bandgap material. This restricts the motion of the electrons and holes, forcing energies for motion to be quantized and only occur at discrete energies.

MUSE Abbreviation for multiple sub-nyquist encoder. A high-definition standard developed in Europe that delivers 1125 lines at 60 frames per second.

mV Abbreviation for millivolt. One thousandth of a Volt or 10^{-3} Volts.

mW Abbreviation for milliwatt. One thousandth of a Watt or 10^{-3} Watts.

MZ Abbreviation for Mach-Zehnder, a structure used in fiber Bragg gratings and interferometers. Named for the men who developed the underlying principles of the structure.

n Abbreviation for nano. One billionth or 10^{-9}.

N Abbreviation for Newtons. Measure of force generally used to specify fiber optic cable strength.

nA Abbreviation for nanoamp. One billionth of an Amp or 10^{-9} Amps.

NA See numerical aperture.

NAB Abbreviation for the National Association of Broadcasters.

NA Mismatch Loss The loss of power at a joint that occurs when the transmitting half has a numerical aperture greater than the NA of the receiving half. The loss occurs when coupling light from a source to fiber, from fiber to fiber, or from fiber to detector.

National Electric Code (NEC) A standard governing the use of electrical wire, cable and fixtures installed in buildings; developed by the NEC Committee of the American National Standards Institute (ANSI), sponsored by the National Fire Protection Association (NFPA), identified by the description ANSI/NFPA 70-1990.

NCTA (National Cable Television Association) The major trade association for the cable television industry.

NDSF See non dispersion-shifted fiber.

Near-end Crosstalk (NEXT, RN) The optical power reflected from one or more input ports, back to another input port. Also known as isolation directivity.

Near Infrared The part of the infrared near the visible spectrum, typically 700 nm to 1500 nm or 2000 nm; it is not rigidly defined.

NEMA Abbreviation for National Electrical Manufacturers Association.

NEP See noise equivalent power.

Network 1. An interconnection of three or more communicating entities and (usually) one or more nodes. 2. A combination of passive or active electronic components that serves a given purpose.

NF See noise figure.

NFPA Abbreviation for National Fire Protection Association.

nm Abbreviation for nanometer. One billionth of a meter or 10^{-9} meters.

Noise Equivalent Power (NEP) The noise of optical receivers, or of an entire transmission system, is often expressed in terms of noise equivalent optical power.

Noise Figure (NF) The ratio of the output signal-to-noise ratio to the input signal-to-noise ratio for a given element in a transmission system. Used for optical and electrical components.

Non Dispersion-shifted Fiber (NDSF) The most popular type of single-mode fiber deployed. It is designed to have zero dispersion near 1310 nm.

Nonlinearity The deviation from linearity in an electronic circuit, an electro-optic device or a fiber that generates undesired components in a signal.

Non Zero-dispersion-shifted Fiber (NZ-DSF) A dispersion shifted SM fiber that has the zero dispersion point near the 1550 nm window, but outside the window actually used to transmit signals. This is a strategy to maximize bandwidth while minimizing fiber nonlinearities.

NRZ Abbreviation for nonreturn to zero. A common means of encoding data that has two states termed "zero" and "one" and no neutral or rest position.

ns Abbreviation for nanosecond. One billionth of a second or 10^{-9} seconds.

NTSC 1. National Television Systems Committee. The organization which formulated the NTSC system. 2. Standard used in the U.S. that delivers 525 lines at 60 frames per second.

Numerical Aperture (NA) The light-gathering ability of a fiber; the maximum angle to the fiber axis at which light will be accepted and propagated through the fiber. The measure of the light-acceptance angle of an optical fiber. $NA = \sin \alpha$, where α is the acceptance angle. NA is also used to describe the angular spread of light from a central axis, as in exiting a fiber, emitting from a source, or entering a detector.

$$\text{Eq. B.7} \qquad NA = \sin\alpha = \sqrt{n_1^2 - n_2^2}$$

Where:

α = Full acceptance angle.
n_1 = Core refractive index.
n_2 = Cladding refractive index.

nW Abbreviation for nanowatt. One billionth of a Watt or 10^{-9} Watts.

NZ-DSF See non zero-dispersion-shifted fiber.

OADM See optical add/drop multiplexer.

OAM Abbreviation for operation, administration, and maintenance. Refers to telecommunications networks.

OCH See optical channel.

OC-x Abbreviation for optical carrier. A carrier rate specified in the SONET standard.

ODN Abbreviation for optical distribution network. Term for optical networks being developed for interactive video, audio, and data distribution.

O/E Abbreviation for optical-to-electrical converter. A device used to convert optical signals to electrical signals. Also OEC.

OEIC Abbreviation for opto-electronic integrated circuit. An integrated circuit that includes both optical and electrical elements.

OEM Abbreviation for original equipment manufacturer. The manufacturer of any device that built to be distributed under the label of another company.

Ohm A unit of electrical resistance equal to the resistance of a circuit in which a potential difference of one Volt produces a current of one Amp.

OLT Abbreviation for optical line termination. Optical network elements that terminate a line signal.

OLTS Optical loss test set. A device used to measure optical loss.

OMD See optical modulation depth.

OMS Abbreviation for optical multiplex section. A section of a DWDM system that incorporates an optical add/drop multiplexer.

1U Abbreviation for "One Unit (U)". U = 1.75 inches.

ONI Abbreviation for optical network interface. A device used in an optical distribution network to connect two parts of that network.

ONU Abbreviation for optical network unit. A network element that is part of a fiber-in-the-loop system interfacing the customer analog access cables and the fiber facilities.

OOI Abbreviation for open optical interface. A point at which an optical signal is passed from one equipment medium to another without conversion to an electrical signal.

Open Systems Interconnect A seven-layer model defined by ISO for defining a communication network.

Optical Add/Drop Multiplexer A device which adds or drops individual wavelengths from a DWDM system.

Optical Amplifier A device that amplifies an input optical signal without converting it into electrical form. The best developed are optical fibers doped with the rare earth element, erbium.

Optical Bandpass The range of optical wavelengths that can be transmitted through a component.

Optical Channel An optical wavelength band for WDM optical communications.

Optical Channel Spacing The wavelength separation between adjacent WDM channels.

Optical Channel Width The optical wavelength range of a channel.

Optical Continuous Wave Reflectometer (OCWR) An instrument used to characterize a fiber optic link wherein an unmodulated signal is transmitted through the link, and the resulting light scattered and reflected back to the input is measured. Useful in estimating component reflectance and link optical return loss.

Optical Directional Coupler (ODC) A component used to combine and separate optical power.

Optical Fall Time The time interval for the falling edge of an optical pulse to transition from 90% to 10% of the pulse amplitude. Alternatively, values of 80% and 20% may be used.

Optical Fiber A glass or plastic fiber that has the ability to guide light along its axis.

Optical Isolator A component used to block out reflected and unwanted light. Used in laser modules, for example. Also called an isolator.

Optical Link Loss Budget The range of optical loss over which a fiber optic link will operate and meet all specifications. The loss is relative to the transmitter output power.

Optical Loss Test Set (OLTS) A source and power meter combined to measure attenuation.

Optical Modulation Depth (OMD) See modulation index.

Optical Path Power Penalty The additional loss budget required to account for degradations due to reflections, and the combined effects of dispersion resulting from intersymbol interference, mode-partition noise, and laser chirp.

Optical Power Meter An instrument that measures the amount of optical power present at the end of a fiber or cable.

Optical Pump Laser A shorter wavelength laser that is used to pump a length of fiber with energy to provide amplification at one or more longer wavelengths. *See also* EDFA.

Optical Return Loss (ORL) The ratio (expressed in units of dB) of optical power reflected by a component or an assembly to the optical power incident on a component port when that component or assembly is introduced into a link or system.

Optical Rise Time The time interval for the rising edge of an optical pulse to transition from 10% to 90% of the pulse amplitude. Alternatively, values of 20% and 80% may be used.

Optical Signal-to-Noise Ratio (OSNR) The optical equivalent of SNR.

Optical Spectrum Analyzer (OSA) A device that allows the details of a region of an optical spectrum to be resolved. Commonly used to diagnose DWDM systems.

Optical Time-division Multiplexing (OTDM) See time-division multiplexing.

Optical Time Domain Reflectometer (OTDR) An instrument that locates faults in optical fibers or infers attenuation by backscattered light measurements.

Optical Waveguide *Synonym* optical fiber.

OSA See optical spectrum analyzer.

OSI See open systems interconnect.

OSNR See optical signal-to-noise ratio.

OSP See outside plant.

OTDM See optical time-division multiplexing.

OTDR See optical time domain reflectometer.

Outside Plant (OSP) 1. In telephony, all cables, conduits, ducts, poles, towers, repeaters, repeater huts, and other equipment located between a demarcation point in a switching facility and a demarcation point in another switching facility or customer premises. 2. In Department of Defense communications, the portion of intrabase communications equipment between the main distribution frame (MDF) and a user end instrument or the terminal connection for a user instrument.

OXC Abbreviation for optical cross-connect. See cross-connect.

p Abbreviation for pico. One trillionth or 10^{-12}.

pA Abbreviation for picoamp. One trillionth of an Amp or 10^{-12} Amps.

PABX Abbreviation for private automatic branch exchange. *Synonym* PBX.

PAL Phase alternation line. A composite color standard used in many parts of the world for TV broadcast. The phase alternation makes the signal relatively immune to certain distortions (compared to NTSC). Delivers 625 lines at 50 frames per second. PAL-plus is an enhanced-definition version.

Parity A term used in binary communication systems to indicate whether the number of 1's in a transmission is even or odd. If the number of 1's is an even number, then parity is said to be even, if the number of 1's is odd, the parity is said to be odd.

Passband The region of frequency in electronics or wavelength in optics which is useful.

Passive Branching Device A device that divides an optical input into two or more optical outputs.

Passive Device Any device that does not require a source of energy for its operation. Examples include electrical resistors or capacitors, diodes, optical fiber, cable, wires, glass lenses and filters.

PBX Abbreviation for private branch exchange. A subscriber-owned telecommunications exchange that usually includes access to public switched networks.

PC Physical contact. Refers to an optical connector that allows the fiber ends to physically touch. Used to minimize backreflection and insertion loss.

PCB Abbreviation for printed circuit board.

PCM See pulse-code modulation.

PCS Fiber See plastic clad silica.

PD See photodiode.

Peak Power Output The output power averaged over that cycle of an electromagnetic wave having the maximum peak value that can occur under any combination of signals transmitted.

Peak Wavelength In optical emitters, the spectral line having the greatest output power. Also called peak emission wavelength.

PFM Pulse-frequency modulation. Also referred to as square wave FM.

Phase Constant The imaginary part of the axial propagation constant for a particular mode, usually expressed in radians per unit length. *See also* attenuation.

Phase Noise Rapid, short-term, random fluctuations in the phase of a wave caused by time-domain instabilities in an oscillator.

Phase-shift Keying (PSK) 1. In digital transmission, angle modulation in which the phase of the carrier is discretely varied in relation either to a reference phase or to the phase of the immediately preceding signal element, in accordance with data being transmitted. 2. In a communications system, the representing of characters, such as bits or quaternary digits, by a shift in the phase of an electromagnetic carrier wave with respect to a reference, by an amount corresponding to the symbol being encoded. Also called biphase modulation, phase-shift signaling.

Photoconductive Losing an electrical charge on exposure to light.

Photodetector An optoelectronic transducer such as a PIN photodiode or avalanche photodiode.

Photodiode (PD) A semiconductor device that converts light to electrical current.

Photon A quantum of electromagnetic energy. A particle of light.

Photonic A term coined for devices that work using photons, analogous to "electronic" for devices working with electrons.

Photovoltaic Providing an electric current under the influence of light or similar radiation.

Pigtail A short optical fiber permanently attached to a source, detector or other fiber optic device.

PINFET PIN detector plus a FET amplifier. Offers superior performance over a PIN alone.

PIN Photodiode See photodiode.

Planer Waveguide A dielectric waveguide fabricated in a flat material such as thin film, that may be used in optical circuits.

Plastic Clad Silica (PCS) Also called hard clad silica (HCS). A step-index fiber with a glass core and plastic or polymer cladding instead of glass.

Plastic Fiber An optical fiber having a plastic core and plastic cladding.

PLC Abbreviation for planar light wave circuit. A device which incorporates a planar waveguide.

Plenum The air handling space between walls, under structural floors, and above drop ceilings, which can be used to route intrabuilding cabling.

Plenum Cable A cable whose flammability and smoke characteristics allow it to be routed in a plenum area without being enclosed in a conduit.

PMD See polarization mode dispersion.

Point-to-point Transmission Transmission between two designated stations.

Polarization The direction of the electric field in the lightwave.

Polarization-maintaining Fiber Fiber that maintains the polarization of light that enters it.

Polarization Mode Dispersion (PMD) Polarization mode dispersion is an inherent property of all optical media. It is caused by the difference in the propagation velocities of light in the orthogonal principal polarization states of the transmission medium. The net effect is that if an optical pulse contains both polarization components, then the different polarization components will travel at different speeds and arrive at different times, smearing the received optical signal.

PON Abbreviation for passive optical network.

Port Hardware entity at each end of the link.

POS Abbreviation for point of sale. The time and place in which a transaction is made.

POTS Abbreviation for plain old telephone service. A call that requires nothing more than basic call handling without additional features.

p-p Abbreviation for peak-to-peak, the algebraic difference between extreme values of a varying quantity.

PPM Abbreviation for pulse-position modulation. A method of encoding data.

Preform The glass rod from which optical fiber is drawn.

Profile Dispersion Dispersion attributed to the variation of refractive index contrast with wavelength.

ps Abbreviation for picosecond. One trillionth of a second or 10^{-12} seconds.

PSK See phase-shift keying.

PSTN Abbreviation for public switched telephone network. A domestic telecommunications network usually accessed by telephones, key telephone systems, private branch exchange trunks and data arrangements.

Pulse A current or voltage which changes abruptly from one value to another and back to the original value in a finite length of time. Used to describe one particular variation in a series of wave motions.

Pulse-code Modulation (PCM) A technique in which an analog signal, such as a voice, is converted into a digital signal by sampling the signal's amplitude and expressing the different amplitudes as a binary number. The sampling rate must be at least twice the highest frequency in the signal.

Pulse Dispersion The spreading out of pulses as they travel along an optical fiber.

Pulse Spreading The dispersion of an optical signal as it propagates through an optical fiber.

Pump Laser A power source for signal amplification, typically a 980 nm or 1480 nm laser, used in EDFA applications.

pW Abbreviation for picowatt. One trillionth of a Watt or 10^{-12} Watts.

QAM See quadrature amplitude modulation.

QDST Abbreviation for quaternary dispersion supported transmission. See DST.

QoS See quality of service.

QPSK See quadrature phase-shift keying.

Quadrature Amplitude Modulation (QAM) A coding technique that uses many discrete digital levels to transmit data with minimum bandwidth. QAM256 uses 256 discrete levels to transmit digitized video.

Quadrature Phase-shift Keying (QPSK) Phase-shift keying uses four different phase angles out of phase by 90°. Also called quadriphase or quaternary phase-shift keying.

Quality of Service (QoS) 1. The performance specification of a communications channel or system which may be quantitatively indicated by channel or system performance parameters, such as signal-to-noise ratio (S/N), bit error ratio (BER), message throughput rate, and call blocking probability. 2. A subjective rating of telephone communications quality in which listeners judge transmissions by qualifiers, such as excellent, good, fair, poor, or unsatisfactory.

Quantum Efficiency In a photodiode, the ratio of primary carriers (electron-hole pairs) created to incident photons. A quantum efficiency of 70% means seven out of ten incident photons create a carrier.

Quaternary Signal A digital signal having four significant conditions. *See also* signal.

Radiation-hardened Fiber An optical fiber made with core and cladding materials that are designed to recover their intrinsic value of attenuation coefficient, within an acceptable time period, after exposure to a radiation pulse.

Radiometer An instrument, distinct from a photometer, to measure power (Watts) of electromagnetic radiation.

Radiometry The science of radiation measurement.

Raman Amplifier An optical amplifier based on the Raman scattering, the generation of many different wavelengths of light from a nominally single-wavelength source by means of lasing action or by the beating together of two frequencies. The optical signal can be amplified by collecting the Raman scattered light.

Rayleigh Scattering The scattering of light that results from small inhomogeneities of material density or composition.

Rays Lines that represent the path taken by light.

Receiver A terminal device that includes a detector and signal processing electronics. It functions as an optical-to-electrical converter.

Receiver Overload The maximum acceptable value of average received power for an acceptable BER or performance.

Receiver Sensitivity The minimum acceptable value of received power needed to achieve an acceptable BER or performance. It takes into account power penalties caused by use of a transmitter with worst-case values of extinction ratio, jitter, pulse rise and fall times, optical return loss, receiver connector degradations, and measurement tolerances. The receiver sensitivity does not include power penalties associated with dispersion, jitter, or reflections from the optical path; these effects are specified separately in the allocation of maximum optical path penalty. Sensitivity usually takes into account worst-case operating and end-of-life (EOL) conditions.

Recombination Combination of an electron and a hole in a semiconductor that releases energy, sometimes leading to light emission.

Refraction The changing of direction of a wavefront in passing through a boundary between two dissimilar media, or in a graded-index medium where refractive index is a continuous function of position.

Refractive Index A property of optical materials that relates to the speed of light in the material.

Refractive Index Gradient The change in refractive index with distance from the axis of an optical fiber.

Refractive Index Profile The description of the value of the refractive index as a function of distance from the optical axis along an optical fiber diameter.

Regenerative Repeater A repeater, designed for digital transmission, in which digital signals are amplified, reshaped, retimed, and retransmitted.

Regenerator *Synonym* regenerative repeater.

Repeater A receiver and transmitter set designed to regenerate attenuated signals. Used to extend operating range.

Residual Loss The loss of the attenuator at the minimum setting of the attenuator.

Responsivity The ratio of a photodetector's electrical output to its optical input in Amperes/Watt (A/W).

Return Loss See optical return loss.

RF Abbreviation for radio frequency. Any frequency within the electromagnetic spectrum normally associated with radio wave propagation.

RFI Abbreviation for radio frequency interference. *Synonym* electromagnetic interference.

RGB Red, green, and blue. The basic parallel component set in which a signal is used for each primary color; or the related equipment or interconnect formats or standards.

Ribbon Cables Cables in which many parallel fibers are embedded in a plastic material, forming a flat ribbon-like structure.

RIN Abbreviation for relative intensity noise. Often used to quantify the noise characteristics of a laser.

Ring A set of stations wherein information passes sequentially between stations that each examine or copy the information, and finally return it to the originating station.

Ring Network A network topology in which terminals are connected in a point-to-point serial fashion in an unbroken circular configuration.

Rise Time The time taken to make a transition from one state to another, usually measured between the 10% and 90% completion points of the transition. Alternatively the rise time may be specified at the 20% and 80% amplitudes. Shorter or faster rise times require more bandwidth in a transmission channel.

RMS Abbreviation for root mean square. Technique used to measure AC voltages.

RTS Abbreviation for request to send. In a communications network, a signal from a remote receiver to a transmitter for data to be sent to that receiver.

RZ Abbreviation for return to zero. A common means of encoding data that has two information states called "zero" and "one" in which the signal returns to a rest state during a portion of the bit period.

s Abbreviation for second.

SAE Abbreviation for Society of Automotive Engineers.

SAN (Storage Area Network) Connects a group of computers to high-capacity storage devices.

S-Band The wavelength range between 1485 nm and 1520 nm used in some CWDM and DWDM applications.

SBS See stimulated Brillouin scattering.

SC Subscription channel connector. A push-pull type of optical connector that originated in Japan. Features high packing density, low loss, low backreflection, and low cost.

Scattering The change of direction of light rays or photons after striking small particles. It may also be regarded as the diffusion of a light beam caused by the inhomogeneity of the transmitting material.

S-CDMA Abbreviation for synchronous code division multiple access. A synchronized version of CDMA.

SCM Abbreviation for subcarrier multiplexing. The process by which multiple subcarrier signals are combined into one signal.

SDM See space-division multiplexing.

SDTV Abbreviation for standard-definition television. *Synonym* NTSC television transmission.

SECAM Abbreviation for Système Èlectronique Couleur avec Mémoire. A TV standard used in various parts of the world. Delivers 625 lines at 50 frames per second.

Selfoc Lens A trade name used by the Nippon Sheet Glass Company for a graded-index fiber lens; a segment of graded-index fiber made to serve as a lens.

Self-phase Modulation (SPM) A fiber nonlinearity caused by the nonlinear index of refraction of glass. The index of refraction varies with optical power level which causes distortion in the waveform.

Semiconductor Optical Amplifier (SOA) A laser diode without end mirrors coupled to fibers on both ends. Light coming in either fiber is amplified by a single pass through the laser diode. An alternative to EDFA's.

Sensitivity See receiver sensitivity.

SH Abbreviation for short-haul. A classification of video performance under RS-250C. Higher performance than long-haul or medium-haul.

Sheath An outer protective layer of a fiber optic cable.

Shot Noise Noise caused by current fluctuations arising from the discrete nature of electrons.

SI Abbreviation for silicon. An element used to make detectors. Good for short wavelengths only (e.g., < 1000 nm).

Sideband Frequencies distributed above and below the carrier that contain energy resulting from amplitude modulation. The frequencies above the carrier are called upper sidebands, and the frequencies below the carrier are called lower sidebands.

Silica Glass Glass made mostly of silicon dioxide, SiO_2, used in conventional optical fibers.

Signal-to-Noise Ratio (SNR) The ratio of the total signal to the total noise which shows how much higher the signal level is than the level of the noise. A measure of signal quality.

Simplex Single element (e.g., a simplex connector is a single-fiber connector).

Simplex Cable A term sometimes used for a single-fiber cable.

Simplex Transmission Transmission in one direction only.

Single Attachment Concentrator A concentrator that offers one attachment to the FDDI network.

Single-line Laser Synonym for single-longitudinal mode laser.

Single-longitudinal Mode Laser (SLM) An injection laser diode which has a single dominant longitudinal mode. A single-mode laser with a side mode suppression ratio (SMSR) > 25 dB.

Single-mode (SM) Fiber A small-core optical fiber through which only one mode will propagate. The typical diameter is 8-9 microns.

Single-mode Laser Diode (SMLD) Synonym for single-longitudinal mode laser.

Single-mode Optical Loss Test Set (SMOLTS) An optical loss test set for use with single-mode fiber.

SI Units Abbreviation for Système Internationale (in English, International System) units, commonly known as the metric system.

SLED See surface-emitting diode.

SLM See Single-longitudinal mode laser.

SMA A threaded type of optical connector. One of the earliest optical connectors to be widely used. Offers poor repeatability and performance.

Smart Structures Also smart skins. Materials containing sensors (fiber optic or other types) to measure their properties during fabrication and use.

SMD Abbreviation for surface-mount device. See SMT.

SMF Abbreviation for single-mode fiber (also SM).

SMPTE Abbreviation for Society of Motion Picture and Television Engineers.

SMT Abbreviation for surface-mount technology. An electronics manufacturing technique.

S/N See signal-to-noise ratio.

SNR See signal-to-noise ratio.

SOA See semiconductor optical amplifier.

Soliton Pulse An optical pulse having a shape, spectral content, and power level designed to take advantage of nonlinear effects in an optical fiber waveguide, for the purpose of essentially negating dispersion over long distances.

SONET Abbreviation for synchronous optical network transport system. An interface standard widely used by the telecom industry to move data. OC-3 is the lowest current rate (155.5 Mb/s) and OC-768 (39.808 Gb/s) is the highest being contemplated. Valid rates increase by a factor of four from the OC-3 rate up to OC-768.

Source In fiber optics, a transmitting LED or laser diode, or an instrument that injects test signals into fibers.

Space-division Multiplexing (SDM) A misnomer; space-division multiplexing has been improperly applied to the use of multiple physical transmission channels, e.g., twisted pairs or optical fibers, under one sheath.

Span Engineering The process of designing a DWDM transmission span to achieve the required performance based on the fiber type, the transmission distance, amplifier spacing, noise, and power, and the channel count.

Spectral Efficiency The number of data bits per second that can be transmitted in a one Hertz bandwidth range.

Spectral Width A measure of the extent of a spectrum. For a source, the width of wavelengths contained in the output at one half of the wavelength of peak power. Typical spectral widths are 50 to 160 nm for an LED and 0.1-5 nm for a laser diode.

Spectral Width, Full Width, Half Maximum (FWHM) The absolute difference between the wavelengths at which the spectral radiant intensity is 50 percent of the maximum power.

Splice A permanent connection of two optical fibers through fusion or mechanical means.

Splitting Ratio The ratio of power emerging from two output ports of a coupler.

SPM See self-phase modulation.

SRS See stimulated Raman scattering.

ST Straight tip connector. Popular fiber optic connector originally developed by AT&T.

Stabilized Light Source An LED or laser diode that emits light with a controlled and constant spectral width, central wavelength, and peak power with respect to time and temperature.

Star Coupler A coupler in which power at any input port is distributed to all output ports.

Star Network A network in which all terminals are connected through a single point, such as a star coupler or concentrator.

STB Abbreviation for set-top box. An auxiliary device that usually sits on top of or adjacent to a television receiver used in direct analog or digital satellite transmission and digital television to view the signals on an analog TV. Also called a set-top converter.

Step-index Fiber Fiber that has a uniform index of refraction throughout the core.

Stimulated Brillouin Scattering (SBS) The easiest fiber nonlinearity to trigger. When a powerful light wave travels through a fiber it interacts with acoustical vibration modes in the glass. This causes a scattering mechanism to be formed that reflects much of the light back to the source.

Stimulated Raman Scattering (SRS) A fiber nonlinearity similar to SBS but having a much higher threshold. This mechanism can also cause power to be robbed from shorter wavelength signals and provide gain to longer wavelength signals.

Strength Member The part of a fiber optic cable composed of aramid yarn, steel strands, or fiberglass filaments that increase the tensile strength of the cable.

Submarine Cable A cable designed to be laid underwater.

Subscriber Loop Also called local loop. The link from the telephone company central office (CO) to the home or business (customer premises).

Supertrunk A cable that carries several video channels between facilities of a cable television company.

Surface-emitting Diode An LED that emits light from its flat surface rather than its side. Simple and inexpensive, with emission spread over a wide angle.

Sync This signal is derived from the composite or combination of the horizontal and vertical drives. *See also* composite sync.

Synchronous A data signal that is sent along with a clock signal.

T Abbreviation for tera. One trillion or 10^{12}.

Tap Loss In a fiber optic coupler, the ratio of power at the tap port to the power at the input port.

Tap Port In a coupler where the splitting ratio between output ports is not equal, the output port containing the lesser power.

TAXI Abbreviation for transparent asynchronous transmitter-receiver interface. A chip used to transmit parallel data over a serial interface.

TBC Abbreviation for timebase corrector, which takes a video signal from a video tape recorder (VTR) and locks it to an external sync signal. The Video Toaster requires that all incoming video from VTR's be timebase corrected.

T-Carrier Generic designator for any of several digitally multiplexed telecommunications carrier systems.

TCP/IP Abbreviation for transmission control protocol/Internet protocol. Two interrelated protocols that are part of the Internet protocol suite. TCP operates on the OSI Transport Layer and breaks data into packets. IP operates on the OSI Network Layer and routes packets. Originally developed by the U.S. Department of Defense.

TDM See time-division multiplexing.

TDMA See time-division multiple access.

TEC Abbreviation for thermoelectric cooler. A device used to dissipate heat in electronic assemblies.

Tee Coupler A three-port optical coupler.

10BASE-F A fiber optic version of an IEEE 802.3 network.

10BASE-FB The portion of 10BASE-F that defines the requirements for a fiber backbone.

10BASE-FL The portion of 10BASE-F that defines a fiber optic link between a concentrator and a station.

10BASE-FP The portion of 10BASE-F that defines a passive star coupler.

10BASE-T A twisted-pair cable version of an IEEE 802.3 network.

10BASE-2 A thin-coaxial-cable version of an IEEE 802.3 network.

10BASE-5 A thick-coaxial-cable version of an IEEE 802.3 network; very similar to the original Ethernet specification.

Telecommunications Management Network (TMN) A network that interfaces with a telecommunications network at several points in order to receive information from, and to control the operation of, the telecommunications network.

Ternary A semiconductor compound made of three elements (e.g., GaAlAs).

TFOCA Tactical fiber optic cable assembly. A rugged fiber optic cable, developed by AT&T, designed for environmentally harsh military applications.

Thermal Noise Noise resulting from thermally induced random fluctuation in current in the receiver's load resistance.

Throughput Loss In a fiber optic coupler, the ratio of power at the throughput port to the power at the input port.

Throughput Port In a coupler where the splitting ratio between output ports is not equal, the output port containing the greater power.

TICL Abbreviation for temperature induced cable loss. Cable loss as a result of extreme temperatures outside a cable's environmental specifications.

Tight-buffer A material tightly surrounding a fiber in a cable, holding it rigidly in place.

Time-division Multiplexing (TDM) A transmission technique whereby several low-speed channels are multiplexed into a high-speed channel for transmission. Each low-speed channel is allocated a specific position based on time.

Time-division Multiple Access (TDMA) A communications technique that uses a common channel (multipoint or broadcast) for communications among multiple users by allocating unique time slots to different users. Used extensively in satellite systems, local area networks, physical security systems, and combat-net radio systems.

TMN See telecommunications management network.

Token Ring A ring-based network scheme in which a token is used to control access to a network. Used by IEEE 802.5 and FDDI.

Total Internal Reflection The reflection that occurs when light strikes an interface at an angle of incidence (with respect to the normal) greater than the critical angle.

TP See twisted pair cable.

Transceiver A device that performs, within one chassis, both telecommunication transmitting and receiving functions.

Transducer A device for converting energy from one form to another, such as optical energy to electrical energy.

Transmitter A device that includes a source and driving electronics. It functions as an electrical-to-optical converter.

Transponder The part of a satellite that receives and transmits a signal.

Tree A physical topology consisting of a hierarchy of master-slave connections between a concentrator and other FDDI nodes (including subordinate concentrators).

Trunk A physical loop topology, either open or closed, employing two optical fiber signal paths, one in each direction (i.e. counter-rotating) forming a sequence of peer connections between FDDI nodes. When the trunk forms a closed loop, it is sometimes called a trunk ring.

TTL Abbreviation for transistor-transistor logic. A type of data format.

Twisted Pair (TP) Cable A cable made up of one or more separately insulated twisted-wire pairs, none of which is arranged with another to form quads.

UL Abbreviation for Underwriter's Laboratory. An organization that test product safety for a wide variety of products.

Unidirectional Operating in one direction only.

UV Abbreviation for ultraviolet. The portion of the electromagnetic spectrum in which the longest wavelength is just below the visible spectrum, extending from approximately 4 nm to approximately 400 nm.

V Abbreviation for Volt. A unit of electrical force or potential, equal to the force that will cause a current of one Amp to flow through a conductor with a resistance of one Ohm.

VAC Abbreviation for Volts, AC. Voltage using alternating current.

VCSEL See vertical cavity surface emitting laser.

VDC Abbreviation for Volts, DC. Voltage using direct current.

VDSL Abbreviation for very high data rate digital subscriber line. A DSL operating at a data rate higher than those of HDSL. *See also* DSL.

Vertical Cavity Surface Emitting Laser Lasers that emit light perpendicular to the plane of the wafer they are grown on. They have very small dimensions compared to conventional lasers and are very efficient.

Vestigial-sideband (VSB) Transmission A modified double-sideband transmission in which one sideband, the carrier, and only a portion of the other sideband are transmitted. *See also* sideband.

VGA Video graphics array. A high-resolution color standard for computer monitors.

Video on Demand (VOD) A term used for interactive or customized video delivery service.

Videoconferencing Conducting conferences via a video telecommunications system.

Videophone A telephone-like service with a picture as well as sound.

Visible Light Electromagnetic radiation visible to the human eye; wavelengths of 400-700 nm.

VOD See video on demand.

Voice Circuit A circuit capable of carrying one telephone conversation or its equivalent; the standard subunit in which telecommunication capacity is counted. The U.S. analog equivalent is 4 kHz. The digital equivalent is 64 kbit/s in North America and in Europe.

VPN Abbreviation for virtual private network. A protected information-system link utilizing tunneling, security controls and end-point address translation giving the end user the impression that a dedicated line exists between nodes.

VSB See vestigial-sideband transmission.

W Abbreviation for Watt. A linear measurement of optical power, usually expressed in milliwatts, microwatts, and nanowatts.

WAN (Wide Area Network) A physical or logical network that provides capabilities for a number of independent devices to communicate with each other over a common transmission-interconnected topology in geographic areas larger than those served by local area networks.

Waveguide A material medium that confines and guides a propagating electromagnetic wave. In the microwave regime, a waveguide normally consists of a hollow metallic conductor, generally rectangular, elliptical, or circular in cross-section. This type of waveguide may, under certain conditions, contain a solid or gaseous dielectric material. In the optical regime, a waveguide used as a long transmission line consists of a solid dielectric filament (optical fiber), usually circular in cross-section. In integrated optical circuits an optical waveguide may consist of a thin dielectric film. In the RF regime, ionized layers of the stratosphere and the refractive surfaces of the troposphere may also serve as a waveguide.

Waveguide Couplers A coupler in which light is transferred between planar waveguides.

Waveguide Dispersion The part of chromatic dispersion arising from the different speeds light travels in the core and cladding of a single-mode fiber (i.e., from the fiber's waveguide structure).

Wavelength The distance between points of corresponding phase of two consecutive cycles of a wave. The wavelength, is related to the propagation velocity, and the frequency, by:

$$\text{Eq. B.8} \qquad \text{Wavelength} = \frac{\text{Propagation Velocity}}{\text{Frequency}}$$

Wavelength Adapter A device which receives one wavelength and outputs a second wavelength, usually to take a standard signal and convert it to an ITU wavelength.

Wavelength-division Multiplexing (WDM) Sending several signals through one fiber with different wavelengths of light.

Wavelength Isolation A WDM's isolation of a light signal in the desired optical channel from the unwanted optical channels. Also called far-end crosstalk.

Wavelength Routing Switch (WRS) A switch, used in optical networks, that routes wavelengths as required to specific terminals in the network.

Wavelength Selective Coupler A device which couples the pump laser wavelength to the optical fiber while filtering out all other unwanted wavelengths. Used in erbium-doped fiber amplifiers.

Wavelength Stability The maximum deviation of the peak wavelength of an optical source from its average wavelength.

WDM See wavelength-division multiplexing.

Wideband Possessing large bandwidth.

WRS See wavelength routing switch.

WWW (World Wide Web) The collection of billions of graphical pages that heavily utilize HTML to provide access to information. One of the key components of the Internet.

XC See cross-connect.

XGM See cross-gain modulation.

XPM See cross-phase modulation.

X-Series Recommendations Sets of data telecommunications protocols and interfaces defined by ITU.

XT See crosstalk.

Y Coupler A variation on the tee coupler in which input light is split between two channels (typically planar waveguide) that branch out like a Y from the input.

Zero-dispersion Slope In single-mode fiber, the rate of change of dispersion with respect to wavelength, at the fiber's zero-dispersion wavelength.

Zero-dispersion Wavelength (λ_0) In a single-mode optical fiber, the wavelength at which material dispersion and waveguide dispersion cancel one another. The wavelength of maximum bandwidth in the fiber.

Zipcord A two-fiber cable consisting of two single-fiber cables having conjoined jackets. A zipcord cable can be easily divided by slitting and puling the conjoined jackets apart.

FIBER OPTIC SYMBOLS

NATIONAL CABLE TELEVISION ASSOCIATION SYMBOLS

The following symbols were voted by the National Cable Television Association's Engineering Committee to become part of the NCTA Recommended Practice for Measurements on Cable Television Systems.

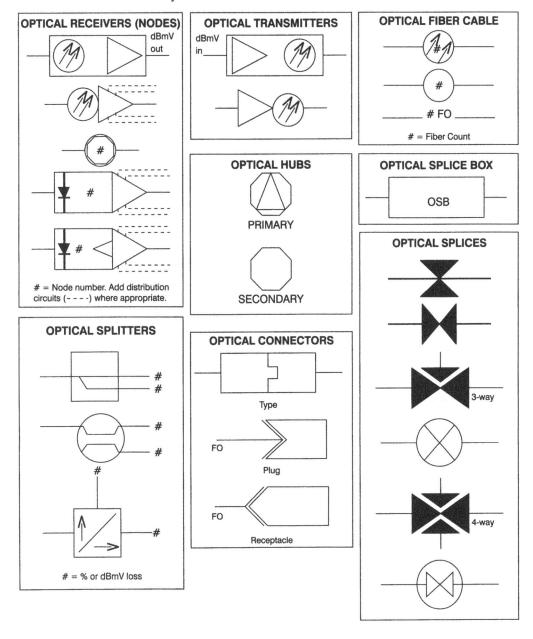

MILITARY SYMBOLS

The following symbols are included in Military Standard MIL-STD-1864A. The purpose is to list the symbols for fiber optic parts for use on military drawings and wherever symbols for fiber optic parts are required.

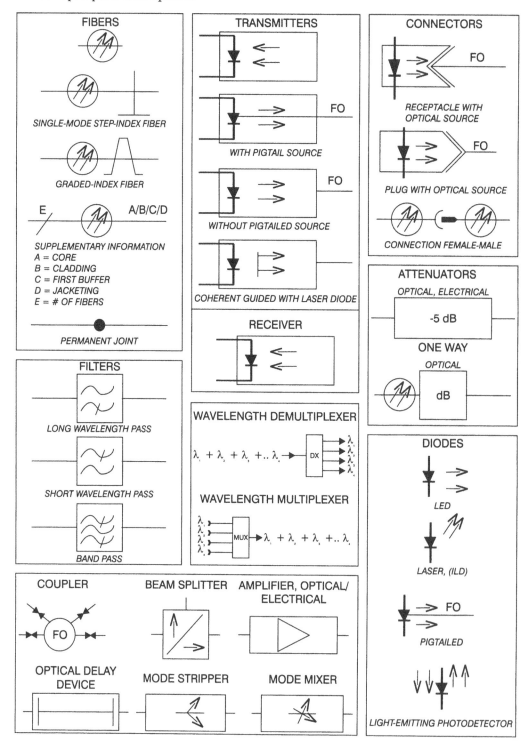

INDUSTRIAL, TELCORDIA & ITU STANDARDS

The following is a listing of the Industrial standards, Telcordia standards (formerly Bellcore standards), and International Telecommunication Union (ITU) standards that apply to the field of fiber optics. Contact information for acquiring copies of these documents is given below.

Industrial Standards
Global Engineering Documents
Attn: Technical Research Department
15 Inverness Way East
Englewood, CO 80112 USA
Phone: 800-624-3974 ext. 1930
            303-792-2181 ext. 1930
Fax: 303-705-4249
E-mail: global.research@ihs.com
Web: http://global.ihs.com

Telcordia Standards
Direct Sales
Telcordia Technologies, Inc
8 Corporate Place, PYA 3A-184
Piscataway, NJ 08854-4156 USA
Phone (US & Canada)
            800-521-2673
Phone (Worldwide)
            732-699-5800
Fax: 732-336-2559
E-mail: telecom-info@telcordia.com
Web: http://www.telcordia.com

ITU Standards
ITU Sales and Marketing Service
Place des Nations
CH-1211
Geneva 20 Switzerland
Phone: 41-22-730-6141
Fax: 41-22-730-5194
E-mail: sales@itu.int
Web: http://www.itu.int

INDUSTRIAL STANDARDS

TSB19	Optical Fiber Digital Transmission Systems: Considerations for Users and Suppliers
TSB62	Informative Test Methods for Fiber Optic Fibers, Cables, Opto-Electronic Sources and Detectors, Sensors, Connecting and Terminating Devices, and Other Fiber Optic Components
TSB62-1	Characterization of Large Flaws in Optical Fibers by Dynamic Tensile Testing with Sensoring Category 3
TSB62-2	Method for Measurement of Hydrogen Evolved from Coated Optical Fiber
TSB63	Reference Guide for Fiber Optic Test Procedures
TSB72	Centralized Optical Fiber Cabling Guidelines
EIA-440-A	Fiber Optic Connector Terminology
TIA/EIA-455-B	Standard Test Procedure for Fiber Optic Fibers, Cables, Transducers, Sensors, Connecting and Terminating Devices, and Other Components
TIA/EIA-455-1B	Cable Flexing for Fiber Optic Interconnecting Devices
TIA/EIA-455-2C	Impact Test Measurements for Fiber Optic Devices
TIA/EIA-455-3A	Procedure to Measure Temperature Cycling Effects on Optical Test Fibers, Optical Cable, and Other Passive Fiber Optic Components
TIA/EIA-455-4B	Fiber Optic Component Temperature Life Test
TIA/EIA-455-5B	Humidity Test Procedure for Fiber Optic Components
EIA/TIA-455-6B	Cable Retention Test Procedure for Fiber Optic Cable Interconnecting Devices
EIA/TIA-455-7	Numerical Aperture of Step-index MM Optical Fibers by Output Far-field Radiation Pattern Measurement
TIA/EIA-455-11B	Vibration Test Procedure for Fiber Optic Components and Cables
EIA/TIA-455-12A	Fluid Immersion Test for Fiber Optic Components
EIA/TIA-455-13A	Visual and Mechanical Inspection of Fibers, Cables, Connectors and/or Other Fiber Optic Devices
EIA/TIA-455-14A	Fiber Optic Shock Test (Specified Pulse)
EIA/TIA-455-15A	Altitude Immersion
EIA/TIA-455-16A	Salt Spray (Corrosion) Test for Fiber Optic Components
EIA-455-20A	Measurement of Change in Optical Transmittance
TIA/EIA-455-21A	Mating Durability for Fiber Optic Interconnecting Devices
EIA/TIA-455-22B	Ambient Light Susceptibility of Fiber Optic Components
EIA-455-23A	Air Leakage Testing of Fiber Optic Components Seals
EIA/TIA-455-24	Water Peak Attenuation Measurement of SM Fibers

EIA/TIA-455-25B	Repeated Impact Testing of Fiber Optic Cables and Cable Assemblies
EIA-455-26A	Crush Resistance of Fiber Optic Interconnecting Devices
TIA/EIA-455-28C	Method for Measuring Dynamic Tensile Strength and Fatigue Parameters of Optical Fibers
EIA/TIA-455-30B	Frequency Domain Measurement of MM Optical Fiber Information Transmission Capacity
TIA/EIA-455-31C	Proof Testing Optical Fibers by Tension
TIA/EIA-455-32A	Fiber Optic Circuit Discontinuities
TIA/EIA-455-33A	Fiber Optic Cable Tensile Loading and Bending Test
TIAEIA-455-34A	Interconnection Device Insertion Loss Test
TIA/EIA-455-35A	Fiber Optic Component Dust (Fine Sand) Test
EIA-455-36A	Twist Test for Fiber Optic Connecting Devices
EIA/TIA-455-37A	Low or High Temperature Bend Test for Fiber Optic Cable
TIA/EIA-455-38	Measurement of Fiber Strain in Cables Under Tensile Load
TIA/EIA-455-39B	Fiber Optic Cable Water Wicking Test
TIA/EIA-455-41A	Compressive Loading Resistance of Fiber Optic Cables
EIA/TIA-455-42A	Optical Crosstalk in Components
TIA/EIA-455-43A	Output Near-field Radiation Pattern Measurement of Optical Waveguide Fibers
TIA/EIA-455-44B	Refractive Index Profile (Refracted Ray Method)
EIA/TIA-455-46A	Spectral Attenuation Measurement for Long-length Graded-index Fibers
EIA/TIA-455-47B	Output Far-field Radiation Pattern Measurement
EIA/TIA-455-48B	Measurement of Optical Fiber Cladding Diameter Using Laser-based Instruments
TIA/EIA-455-49A	Procedure for Measuring Gamma Radiation Effects on Optical Fibers and Optical Cables
TIA/EIA-455-50B	Light Launch Conditions for Long-length Graded-index Optical Fibers Spectral Attenuation Measurements
EIA/TIA-455-51A	Pulse Distortion Measurement of MM Glass Optical Fiber Information Capacity
EIA/TIA-455-53A	Attenuation by Substitution Measurement for MM Graded-index Optical Fibers or Fiber Assemblies Used in Long-length Communications Systems
TIA/EIA-455-54B	Mode Scrambler Requirements for Overfilled Launching Conditions to MM Fibers
TIA/EIA-455-55C	End-view Methods for Measuring Coating and Buffer Geometry of Optical Fibers
TIA/EIA-455-56B	Test Method for Evaluating Fungus Resistance of Optical Waveguide Fibers and Cables
TIA/EIA-455-57B	Preparation and Examination of Optical Fiber Endface for Testing Purposes
EIA/TIA-455-58A	Core Diameter Measurements of Graded-index Optical Fibers
EIA/TIA-455-59	Measurement of Fiber Point Defects Using an OTDR
EIA/TIA-455-60	Measurement of Fiber or Cable Length Using an OTDR
EIA/TIA-455-61	Measurement of Fiber or Cable Attenuation Using an OTDR
EIA/TIA-455-62A	Measurement of Optical Fiber Macrobend Attenuation
TIA/EIA-455-64	Procedure for Measuring Radiation-induced Attenuation in Optical Fibers and Optical Cables
TIA/EIA-455-67	Procedure for Assessing High Temperature Aging Effects on Optical Characteristics of Optical Fibers
EIA/TIA-455-69A	Test Procedure for Evaluating Minimum and Maximum Exposure Temperature on the Optical Performance of Optical Fiber
TIA/EIA-455-70	Procedure for Assessing High Temperature Exposure Effects on Mechanical Characteristics of Optical Fibers
TIA/EIA-455-71A	Procedure to Measure Temperature Shock Effects on Fiber Optic Components
TIA/EIA-455-72	Procedure for Measuring Temperature and Humidity Cycling Exposure Effects on Optical Characteristics of Optical Fibers
TIA/EIA-455-73	Procedure for Measuring Temperature and Humidity Cycling Aging Effects on Mechanical Characteristics of Optical Fibers
TIA/EIA-455-74	Fluid Immersion Aging Procedure for Optical Fiber Optical Properties

TIA/EIA-455-75A	Fluid Immersion Test for Optical Waveguide Fibers
TIA/EIA-455-76	Method for Measuring Dynamic Fatigue of Optical Fibers by Tension
EIA/TIA-455-77	Procedures to Qualify a Higher-order Mode Filter for Measurements of SM Fibers
EIA/TIA-455-78A	Spectral Attenuation Cutback Measurement for SM Optical Fibers
TIA/EIA-455-80B	Measurement of Cutoff Wavelength of SM Fiber by Transmitted Power
EIA/TIA-455-81A	Compound Flow (Drip) Test for Filled Fiber Optic Cable
EIA/TIA-455-82B	Fluid Penetration Test for Fluid-Blocked Fiber Optic Cable
TIA/EIA-455-84B	Jacket Self-adhesion (Blocking) Test for Fiber Optic Cable
TIA/EIA-455-85A	Fiber Optic Cable Twist Test
EIA-455-86	Fiber Optic Cable Jacket Shrinkage
TIA/EIA-455-87B	Fiber Optic Cable Knot Test
TIA/EIA-455-88	Fiber Optic Cable Bend Test
TIA/EIA-455-89B	Fiber Optic Cable Jacket Elongation and Tensile Strength Test
TIA/EIA-455-91	Fiber Optic Cable Twist-Bend Test
TIA/EIA-455-92	Optical Fiber Cladding Diameter and Noncircularity Measurement by Fizeau Interferometry
TIA/EIA-455-93	Test Method for Optical Fiber Cladding Diameter and Non-circularity by Noncontacting Michelson Interferometry
TIA/EIA-455-95	Absolute Optical Power Test for Optical Fibers and Cables
EIA/TIA-455-98A	Fiber Optic Cable External Freezing Test
TIA/EIA-455-99	Gas Flame Test for Special Purpose Fiber Optic Cable
TIA/EIA-455-100A	Gas Leakage Test for Gas-blocked Fiber Optic Cable
TIA/EIA-455-104A	Fiber Optic Cable Cyclic Flexing Test
TIA/EIA-455-106	Procedure for Measuring the Near Infrared Absorbance of Fiber Optic Coating Materials
TIA/EIA-455-107-A	Determination of Component Reflectance or Link/System Return Loss Using a Loss Test Set
TIA/EIA-455-113	Polarization Mode Dispersion Measurement for SM Optical Fibers by Wavelength Scanning
TIA/EIA-455-115	Spectral Attenuation Measurement of Step-index MM Optical Fibers
TIA/EIA455-119	Coating Geometry Measurement for SM Optical Fibers by Interferometry
TIA/EIA-455-120	Modeling Spectral Attenuation on Optical Fiber
TIA/EIA-455-122	Polarization Mode Dispersion Measurement for SM Optical Fibers by Jones Matrix Eigenanalysis
TIA/EIA-455-124	Polarization-mode Dispersion Measurement for SM Optical Fibers by Interferometry
EIA/TIA-455-127	Spectral Characterization of MM Laser Diodes, Performance of Optical Fibers
TIA/EIA-455-131	Measurement of Optical Fiber Ribbon Residual Twist
TIA/EIA-132	Measurement of the Effective Area of SM Optical Fiber
TIA/EIA-133	Length Measurement of Polarization Dependent Loss (PDL) of SM Fiber Optic Components
TIA/EIA-455-141	Twist Test for Optical Fiber Ribbons
TIA/EIA-455-157	Measurement of Polarization Dependent Loss (PDL) of SM Fiber Optic Components
TIA/EIA-455-158	Measurement of Breakaway Frictional Force in Fiber Optic Connector Alignment Sleeves
TIA/EIA-455-160	Procedure for Assessing Temperature and Humidity Exposure Effects on Optical Characteristics of Optical Fiber
TIA/EIA-455-161	Procedure for Assessing Temperature and Humidity Exposure Effects on Mechanical Characteristics of Optical Fibers
TIA/EIA-455-162A	Fiber Optic Cable Temperature-Humidity Cycling
EIA/TIA-455-164A	SM Fiber, Measurement of Mode Field Diameter by Far-field Scanning
EIA/TIA-455-165A	Mode Field Diameter Measurement by Near-field Scanning (SM)
EIA/TIA-455-167A	Mode Field Diameter Measurement, Variable Aperture Method in Far-field
EIA/TIA-455-168-A	Chromatic Dispersion Measurement of MM Graded-index and SM Optical Fibers by Spectral Group Delay Measurement in the Time Domain
TIA/EIA-455-169A	Chromatic Dispersion Measurement of Optical Fibers by Phase-Shift Method
EIA/TIA-455-170	Cable Cutoff Wavelength of SM Fiber by Transmitted Power

TIA/EIA-455-171	Attenuation by Substitution Measurement Short-length MM Graded-index and SM Fiber Optic Cable Assemblies
TIA/EIA-455-172	Flame Resistance of Firewall Connector
EIA/TIA-455-173	Coating Geometry Measurement of Optical Fiber, Side-view Method
TIA/EIA-455-175A	Chromatic Dispersion Measurement of SM Optical Fibers by the Differential Phase-shift Method
TIA/EIA-455-176	Method for Measuring Optical Fiber Cross-sectional Geometry by Automated Grayscale Analysis
TIA/EIA-455-177A	Numerical Aperture Measurement of Graded-index Optical Fibers
TIA/EIA-455-178A	Measurements of Strip Force Required for Mechanically Removing Coatings from Optical Fibers
TIA/EIA-455-179	Inspection of Cleaved Fiber End Faces by Interferometry
TIA/EIA-455-180	Measurement of Optical Transfer Coefficients of a Passive Branching Device
TIA/EIA-455-181	Lightning Damage Susceptibility Test for Fiber Optic Cables with Metallic Components
TIA/EIA-455-184	Coupling Proof Overload Test for Fiber Optic Interconnecting Devices
TIA/EIA-455-185	Strength of Coupling Mechanism for Fiber Optic Interconnecting Devices
TIA/EIA-455-186	Gauge Retention Force Measurement for Components
TIA/EIA-455-187	Engagement and Separation Force of Fiber Optic Connector Sets
EIA/TIA-455-188	Low-temperature Testing of Fiber Optic Components
EIA/TIA-455-189	Ozone Exposure Test of Fiber Optic Components
EIA/TIA-455-190	Low Air Pressure (High Altitude) Testing of Fiber Optic Components
TIA/EIA-455-191	Measurement of Mode Field Diameter of SM Optical Fiber
TIA/EIA-455-192	H-Parameter Test Method for Polarization-maintaining Optical Fiber
TIA/EIA-455-193	Polarization Crosstalk Test Method for Polarization-maintaining Optical Fiber
TIA/EIA-455-196	Guidelines for Polarization-mode Measurement in SM Fiber Optic Components and Devices

EIA/TIA-458-B	Standard Optical Fiber Material Classes and Preferred Sizes
TIA/EIA-472C000-A	Sectional Specification for Fiber Optic Communications Cables for indoor use.
TIA/EIA-472D000-A	Sectional Specification for Fiber Optic Communication Cables for Outside Plant Use
TIA/EIA-4750000-C	Generic Specification for Fiber Optic Connectors
EIA/TIA-475C000	Sectional Specification for Type FSMA Connectors
EIA/TIA-475CA00	Blank Detail Specification for Optical Fibers and Cable Type FSMA, Environmental Category I
EIA/TIA-475CB00	Blank Detail Specification for Optical Fibers and Cables Type FSMA, Environmental Category II
EIA/TIA-475CC00	Blank Detail Specification Connector Set for Optical Fibers and Cables Type FSMA, Environmental Category III.
TIA/EIA-475EA00	Blank Detail Specification for Connector Set for Optical Fibers and Cables, Type BFOC/2.5, Environmental Category I
TIA/EIA-475EB00	Blank Detail Specification for Connector Set for Optical Fibers and Cables, Type BFOC/2.5, Environmental Category II
TIA/EIA-475EC00	Blank Detail Specification for Connector Set for Optical Fibers and Cables, Type BFOC/2.5, Environmental Category III
TIA/EIA-4920000-A	Generic Specification for Optical Waveguide Fibers
TIA/EIA-492A000-A	Sectional Specification for Class Ia MM, Graded-index Optical Waveguide Fibers
TIA/EIA-492AA00-A	Blank Detail Specification for Class Ia MM, Graded-index Optical Waveguide Fibers
EIA-492B000	Sectional Specification for Class IV SM Optical Waveguide Fibers
EIA-492BA00	Blank Detail Specification for Class IVa Dispersion, Unshifted SM Optical Waveguide Fibers
EIA/TIA-492BB00	Blank Detail Specification for Class IVb Dispersion, Shifted SM Optical Waveguide Fibers
TIA/EIA-492C000	Sectional Specification for Class IVa Dispersion-Unshifted SM Optical Fibers

TIA/EIA-492CA00	Blank Detail Specification for Class IVa Dispersion-Unshielded SM Optical Fibers
TIA/EIA-492CAAA	Detail Specification for Class IVa Dispersion-Unshifted SM Optical Fibers
TIA/EIA-492E000	Sectional Specification for Class IVd Nonzero Dispersion SM Optical Fibers for the 1550 nm Window
TIA/EIA-492EA000	Blank Detail Specification for Class IVd Nonzero Dispersion SM Optical Fiber for the 1550 nm Window
TIA/EIA-5090000	Generic Specification for Fiber Optic Terminal Devices
TIA/EIA-5150000	Generic Specification for Optical Fiber and Cable Splices
TIA/EIA-515B0000	Sectional Specification for Splice Closures for Pressurized Aerial, Buried, and Underground Fiber Optic Cables
TIA/EIA-526	Standard Test Procedures for Fiber Optic Systems
EIA/TIA-526-2	Effective Transmitter Output Power Coupled into SM Fiber Optic Cable
EIA/TIA-526-3	Fiber Optic Terminal Equipment Receiver Sensitivity and Maximum Receiver Input
TIA/EIA-526-7	Measurement of Optical Power Loss of Installed SM Fiber Cable Plant
EIA/TIA-526-10	Measurement of Dispersion Power Penalty in Digital SM Systems
TIA/EIA-526-11	Measurement of Single-reflection Power Penalty for Fiber Optic Terminal Equipment
EIA/TIA-526-14A	Optical Power Loss Measurements of Installed MM Fiber Cable Plant
TIA/EIA-526-15	Jitter Tolerance Measurement
TIA/EIA-526-16	Jitter Transfer Function Measurement
TIA/EIA-526-17	Output Jitter Measurement
TIA/EIA-526-18	Systematic Jitter Generation Measurement
TIA/EIA-5430000	Generic Specification, Field Portable Electronic Instruments for Optical Fiber System Measurements
EIA/TIA-559-1	SM Fiber Optic System Transmission Design
TIA/EIA-5730000-A	Generic Specification for Field Portable Fiber Optic Tools
TIA/EIA-573A000-A	Sectional Specification for Field Portable Optical Fiber Cleaving Tools

TIA/EIA-573B000-A	Sectional Specification for Field Portable Single Optical Fiber Stripping Tools
TIA/EIA-573C000	Sectional Specification for Field Portable Optical Microscopes
TIA/EIA-573D000	Sectional Specification for Field Portable Polishing Devices for Preparation of Optical Fibers
TIA/EIA-587	Fiber Optic Graphic Symbols
TIA/EIA-590-A	Standard for Physical Location and Protection of Belowground Fiber Optic Cable Plant
TIA/EIA-598-A	Optical Fiber Cable Coding
TIA/EIA-604	Fiber Optic Connector Intermateablity Standards
TIA/EIA-620A000	Sectional Specification for SM Fiber Optic Branching Devices for Outside Plant Applications
TIA/EIA-620AA00	Blank Detail Specification for SM Fiber Optic Branching Devices for Outside Plant Applications
TIA/EIA-626	MM Fiber Optic Link Transmission Design
TIA/EIA-6300000	Generic Specification for Passive Fiber Optic Switches

TELCORDIA STANDARDS

GR-20	Generic Requirements of Optical Fiber and Optical Fiber Cable
TA-TSY-00038	Digital Fiber Optic System Requirements and Objectives
ST-TEC-000051	Telecommunications Transmission Engineering Textbook, Volume 1 - Principles, 3rd Edition
ST-TEC-000052	Telecommunications Transmission Engineering Textbook, Volume 2 Facilities, 3rd Edition
ST-TEC-000053	Telecommunications Transmission Engineering, Volume 3 - Networks And Services, 3rd Edition
TR-TSY-000187	Optical Cable Placing Winches
GR-196	Generic Requirements for OTDR Equipment
GR-198	Generic Requirements for Optical Loss Test Sets
TR-NWT-000233	Wideband and Broadband Digital Cross-connect Systems Generic Requirements and Objectives
GR-253	Synchronous Optical Network (SONET) Transport Systems: Common Generic Criteria

TR-NWT-000264	Generic requirements for Optical Fiber Cleaves	GR-909	Generic Requirements and Objectives for Fiber-in-the-loop Systems
GR-326	Generic Requirements for SM Optical Connectors and Jumper Assemblies	GR-910	Generic Requirements for Fiber Optic Attenuators
GR-356	Generic Requirements for Optical Cable Interduct	TR-NWT-000917	SONET Regenerator (SONET RGTR) Equipment Generic Criteria
GR-409	Generic Requirements for Premises Fiber Optic Cable	TR-TSY-000949	Generic Requirements for Service Terminal Closures with Optical Cable
GR-418	Generic Reliability Assurance Requirements for Fiber Optic Transport Systems	GR-950	Generic Requirements for Optical Network Unit (ONU) Closures
TR-TSY-000441	Submarine Splice Closures for Fiber Optic Cable	TR-NWT-001001	Generic Requirements for Cable Shield Bonding Clamps
GR-449	Generic Requirements and Design Considerations for Fiber Distributing Frames	GR-1009	Generic Requirements for Fiber Optic Clip-on Test Sets
TA-NPL-000464	Generic Requirements and Design Considerations for Optical Digital Signal Cross-connect Systems	TR-TSY-001028	Generic Criteria for Optical Continuous Wave Reflectometers
TR-NWT-000468	Reliability Assurance Practices for Optoelectronic Devices in Central Office Applications	TA-TSY-001040	SONET Test Sets for Acceptance and Maintenance Testing: Generic Criteria
GR-496	SONET Add/Drop Multiplexer (SONET ADM) Generic Criteria	GR-1073	Generic Requirements for Fiber Optic Switches
		GR-1073	Generic Requirements for SM Fiber Optic Switches
SR-TSY-000686	Cost Comparison of 45 Mb/s Video Over Fiber with Digital Alternate Technologies	GR-1081	Generic Requirements for Field-mountable Optical Fiber Connectors
GR-761	Generic Criteria for Chromatic Dispersion Test Sets	GR-1095	Generic Requirements for Multi-fiber Splicing Systems for SM Optical Fibers
TR-NWT-000764	Generic Criteria for Optical Fiber Identifiers		
GR-765	Generic Requirements for Single Fiber SM Optical Splices and Splicing Systems	TR-NWT-001121	Generic Requirements for Self-supporting Optical Fiber Cable
GR-769	Generic Requirements for Organizer Assemblies	SR-TSY-001129	Communicating by Light Fiber Optics Slide Package
GR-771	Generic Requirements for Fiber Optic Splice Closures	TR-NWT-001137	Generic Requirements for Hand-held Optical Power Meters
TR-TSV-000774	SMDS Operations Technology Network Element Generic Requirements	SR-TSY-001171	Methods and Procedures for System Reliability Analysis
GR-782	SONET Digital Switch Trunk Interface Criteria	TR-NWT-001190	Generic Requirements for Fiber Optic Cable Locators
		TR-NWT-001196	Generic Requirements for Splice Verification Sets
TR-TSY-000789	Generic Requirements for Lashed Cable Supports		
SR-NWT-000821	Field Reliability Performance Study Handbook	GR-1209	Generic Requirements for Passive Optical Components
TR-TSY-000843	Generic Requirements for Optical and Optical/Metallic Buried Service Cable	GR-1222	Generic Requirements for Fiber Optic Terminators
		GR-1295	Generic Requirements for Remote Fiber Testing Systems
TR-TSY-000886	Generic Criteria for Optical Power Meters	GR-1309	TSC/RTU and OTAU Generic Requirements for Remote Optical Fiber Testing
TR-TSY-000887	Generic Criteria for Fiber Optic Stabilized Light Sources		
TR-TSY-000902	Generic Requirements for Non-Concrete Splice Enclosures	GR-1312	Generic Requirements for Optical Fiber Amplifiers and Proprietary Dense Wavelength-division Multiplexed Systems

TR-NWT-001319	Generic Requirements for Fiber Optic Visual Fault Finders	GR-2918	DWDM Network Transport Systems with Digital Tributaries for Use in Metropolitan Area Applications: Common Generic Criteria
TR-NWT-001322	Generic Requirements for Steam Resistant Cables		
GR-1345	Framework Generic Requirements for Element Manager (EM) Applications for SONET Subnetworks	GR-2919	Generic Requirements for Hybrid Splice/Connector for SM Optical Fiber
		GR-2923	Generic Requirements for Optical Fiber Connector Cleaning Products
SR-TSY-001369	Introduction to Reliability of Laser Diodes and Modules	GR-2947	Generic Requirements for Portable Polarization Mode Dispersion Test Sets
GR-1380	Generic Requirements for Fusion Splice Protectors	
TA-NWT-001416	Generic Requirements for Fiber Optic Talksets	GR-2952	Generic Requirements for Polarization Wavelength-division Multiplexer Analyzers
SR-TSY-001425	Temperature Cycling Test of Laser Modules Used for Interoffice Applications	GR-2961	Generic Requirements for Multi-purpose Fiber Optic Cable
GR-1435	Generic Requirements for Multi-fiber Optical Connectors	GR-2979	Common Generic Requirements for Optical Add/Drop Multiplexers (OADMs) and Optical Terminal Multiplexers (OTMs)
TA-NWT-001500	Generic Requirements for Power Optical Network Units in Fiber-in-the-Loop Systems	
SR-NWT-001756	Automatic Protection Switching For SONET	GR-2998	Generic Requirements for Wavelength-division Multiplexing (WDM) Element Management Systems (EMSs)
SR-NWT-002014	Suggested Optical Cable Code	
SR-NWT-002041	Transport Surveillance for Time-division Multiple Access (TDMA) Based Point-to-multipoint Fiber-in-the-loop Systems	GR-2999	Generic Requirements for Wavelength-division Multiplexing (WDM) Network Management Systems (NMSs)
SR-OPT-002104	TIRKS® Time Slot Numbering Schemes for SONET Add/Drop Multiplex Equipment (ADM)	GR-3105	Generic Requirements for Dense Wavelength-division Multiplexing (DWDM) Network Management System (NMS)-Element Management System (EMS) Interface
SR-ARH-002744	SM Fiber Connectors Technology	
SR-2751	OCS OS-NE Interface Support for the CMISE/OSI Protocol Stack	SR-3244	Reliability Concerns With Lightwave Components
		SR-3317	OPS/INE to Network Element Generic OSI/CMISE Interface Support
TR-NWT-002811	Generic Requirements for Cable Placing Lubricants	
GR-2853	Generic Requirements for AM/Digital Video Laser Transmitters and Receivers	SR-3904	AM/Digital Video Laser Transmitters, of as, and Receivers Certification
GR-2854	Generic Requirements for Fiber Optic Dispersion Compensators	SR-3928	Optical Fiber and Fiber Optic Cable Certification
SR-NWT-002855	Optical Isolators: Reliability Issues and Proposed Tests	SR-4037	Degradation in Transmission Performance of Fiber Optic Video Circuits
GR-2866	Generic Requirements for OPtical Fiber Ribbon Fanouts	SR-4226	Fiber Optic Connector Certification
GR-2882	Generic Requirements for Optical Isolators and Circulators	SR-4263	SM Optical Fiber Splice and Splicing System Certification
GR-2883	Generic Requirements for Fiber Optic Filters	SR-4301	Single and Multi-fiber Fusion Slice Protector Certification
GR-2898	Generic Requirements for Fiber Demarcation Boxes	SR-4588	Test Plan of Critical Parameters for High Density Fiber Distributing Frames
GR-2903	Reliability Assurance Practices for Fiber Optic Data Links	SR-4731	Optical Time Domain Reflectometer Data Format

ITU STANDARDS

G.650	Definition and Test Methods for the Relevant Parameters of SM Fibers
G.651	Characteristics of a 50/125 μm MM Graded-index Optical Fiber Cable
G.652	Characteristics of a SM Optical Fiber Cable
G.653	Characteristics of a Dispersion-shifted SM Optical Fiber Cable
G.654	Characteristics of a Cut-off Shifted SM Optical Fiber Cable
G.655	Characteristics of a Non Zero-dispersion-shifted SM Optical Fiber Cable
G.661	Definition and Test Methods For the Relevant Generic Parameters of Optical Fiber Amplifiers
G.662	Generic Characteristics of Optical Fiber Amplifier Devices and Sub-Systems
G.663	Application Related Aspects of Optical Fiber Amplifier Devices and Sub-systems
G.664	Optical Safety Procedures and Requirements for Optical Transport Systems
G.671	Transmission Characteristics of Passive Optical Components
G. 681	Functional Characteristics of Interoffice and Long-haul Line Systems Using Optical Amplifiers, Including Optical Multiplexing
G.691	Optical Interfaces for Single Channel STM-64, STM-256 and Other SDH Systems with Optical Amplifiers
G.692	Optical Interfaces for Multichannel Systems with Optical Amplifiers
G.709	Network Node Interface for the Optical Transport Network (OTN)
G.871/Y.1301	Framework of Optical Transport Network Recommendations
G.872	Architecture of Optical Transport Networks
G.911	Parameters and Calculation Methodologies for Reliability and Availability of Fiber Optic Systems
G.957	Optical Interfaces for Equipment and Systems Relating to the Synchronous Digital Hierarchy
G.958	Digital Line Systems Based on the Synchronous Digital Hierarchy for Use on Optical Fiber Cables
G.959	Optical Transport Network Physical Layer Interfaces
G.981	PDH Optical Line Systems for the Local Network
G.982	Optical Access Networks to Support Services Up To the ISDN Primary Rate or Equivalent Bit Rates
G.983.1	Broadband Optical Access Systems Based on Passive Optical Networks (PON)
G.983.2	OTN Management and Control Interface Specification for ATM PON
L.10	Optical Fiber Cables for Duct, Tunnel, Aerial and Buried Application
L.12	Optical Fiber Joints
L.13	Sheath Joints and Organizers of Optical Fiber Cables in the Outside Plant
L.14	Measurement Method to Determine the Tensile Performance of Optical Fiber Cables Under Load
L.15	Optical Local Distribution Networks Factors to be Considered For Metal Cable Sheaths
L.25	Optical Fiber Cable Network Maintenance
L.26	Optical Fiber Cables for Aerial Application
L.27	Method for Estimating the Concentration of Hydrogen in Optical Fiber Cables
L.31	Optical Fiber Attenuators
L.34	Installation of Optical Fiber Ground Wire (OGW) Cable
L.35	Installation of Optical Fiber Cable in the Access Network
L.36	SM Fiber Optic Connectors
L.37	Fiber Optic (non-wavelength selective) Branching Devices
L.40	Optical Fiber Outside Plant Maintenance Support, Monitoring, and Testing System
L.41	Maintenance Wavelength on Fibers Carrying Signals
L.47	Access Facilities Using Hybrid Fiber/Copper Networks

SOCIETIES, CONFERENCE SPONSORS & TRADE MAGAZINES

SOCIETIES & CONFERENCE SPONSORS

American Electronics Association

5201 Great American Parkway, Suite 520
Santa Clara, CA 95054
TEL: (408) 987-4200
http://www.aeanet.org

American Institute of Physics (AIP)

One Physics Ellipse
College Park, MD 20740-3843
TEL: (301) 209-3200
http://www.aip.org

American National Standards Institute (ANSI)

25 West 43rd Street
New York, NY 10036
TEL: (212) 642-4900
http://www.ansi.org

American Physical Society (APS)

One Physics Ellipse
College Park, MD 20740-3844
TEL: (301) 209-3200
http://www.aps.org

American Society for Industrial Security (ASIS)

1625 Prince Street
Alexandria, Virginia 22314-2818
TEL: (703) 519-6200
http://www.asisonline.org

Armed Forces Communication and Electronic Association (AFCEA)

4400 Fairfax Lakes Court
Fairfax, VA 22033-3899
TEL: (703) 631-6100
USA: (800) 336-4583
http://www.afcea.org

Building Industry Consulting Service International, Inc. (BICSI)

8610 Hidden River Parkway
Tampa, FL 33637-1000
TEL: (813) 979-1991
USA: (800) 242-7405
http://www.bicsi.org

Electronic Industries Association (EIA)

2500 Wilson Boulevard
Arlington, VA 22201-3834
TEL: (703) 907-7500
http://www.eia.org

ElectroniCast Corp.

800 S. Claremont St.
San Mateo, CA 94402
TEL: (650) 343-1398
http://www.electronicast.com

European Conference on Optical Communication (ECOC)

Nexus Media Limited
Nexus House, Swanley, Kent, BR8 8HU, UK
TEL: (44) 1322-66-00-70
http://www.ecoc-exhibition.com

European Optical Society (EOS)

Klaus-Dieter Nowitzki
Holleritallee 8
30419 Hannover Germany
TEL: 49-0-511-2788-115
http://www.EuropeanOpticalSociety.org

European Physical Society (EPS)

34 rue Marc Seguin, BP 2136
F-68060 Mulhouse Cedex, France
TEL: 33-389-32-94-40
http://www.eps.org

Events Management International, Inc.

100 Ledgewood Place, Suite 203
Rockland, MA 02370
TEL: (781) 871-5600
http://www.netsultants.com

Fiber Optic Association, Inc.

Box 230851
Boston, MA 02123-0851
Tel: 1-617-469-2FOA
http://www.theFOA.org

Hannover Fairs USA, Inc.

103 Carnegie Center
Princeton, NJ 08540
TEL: (609) 987-1202
http://www.hfusa.com

Institute of Electrical Engineers (IEE)

Savory Place
London, WC2R CBL-UK
TEL: 44-0-20-7240-1871
http://www.iee.org

Institute of Electrical and Electronics Engineers (IEEE)

445 Hoes Lane
Piscataway, NJ 0885-1311
TEL: (732) 981-0060
http://www.ieee.org

International Commission for Optics (ICO)

B.P. 147, 91403
Orsay cedex
France
TEL: 33-1-69-35-87-41
http://www.ico-optics.org

International Communications Association (ICA)

PO Box 9589
Austin, TX 78766
TEL: (512) 451-6720
http://www.icahdq.org

International Communications Industries Association

11242 Waples Mill Road, Suite 200
Fairfax, VA 22030
TEL: (703) 273-7200
USA: (800) 659-7469
http://www.infocomm.org

International Engineering Consortium (IEC)

549 W. Randolph, Suite 600
Chicago, IL 60661-2208
TEL: (312) 559-4100
http://www.iec.org

International Society for Optical Engineering (SPIE)
PO Box 10
Bellingham, WA 98227-0010
TEL: (360) 676-3290
http://www.spie.org

International Society For Measurement and Control (ISA)
67 Alexander Drive
PO Box 12277
Research Triangle Park
North Carolina 27709
TEL: (919) 549-8411
http://www.isa.org

International Telecommunication Union (ITU)
Places Des Nations, CH-1211
Geneva, 20, SWITZERLAND
TEL: (44) 22-730-6141
http://www.itu.int

International Wire & Cable Symposium (IWCS)
174 Main Street
Eatontown, NJ 07724
TEL: (732) 389-0990
http://www.iwcs.org

Kallman Worldwide, Inc.
4 North Street
Waldwick, NJ 07463-1842
TEL: (201) 251-2600
http://www.kaliman.com

Kessler Marketing Intelligence (KMI)
The Foundry Corporate Office Center
235 Promenade Street, Suite 400
Providence, RI 02908-5734
USA: (800) 343-5734
TEL: (401) 243-8100
http://www.kmicorp.com

E.J. Krause & Associates, Inc.
6550 Rock Spring Drive, Suite 500
Bethesda, MD 20817
TEL: (301) 493-5500
http://www.ejkrause.com

Lasers and Electro-Optics Society (LEOS)
445 Hoes Lane
PO Box 1331
Piscataway, NJ 08855-1331
TEL: (201) 562-3892
http://www.i-LEOS.org

Laser Institute of America (LIA)
13501 Ingenuity Drive, Suite 128
Orlando, FL 32826
USA: (800) 34-LASER
TEL: (407) 380-1553
http://www.laserinstitute.org

Media Communications Association-International (MCA-I)
9202 N. Meridian Street, Suite 200
Indianapolis, IN 46260-1810
TEL: (317) 816-6269
http://www.itva.org

National Association of Broadcasters (NAB)
1771 N Street., NW
Washington, D.C. 20036
TEL: (202) 429-5300
http://www.nab.org

National Association of State Telecommunications Directors (NASTD)
2760 Research Park
Lexington, KY 40578-1910
TEL: (859) 244-8186
http://www.nastd.org

National Association of Telecommunications Officers & Advisors (NATOA)
1595 Spring Hill Road, Suite 330
Vienna, VA 22182
TEL: (703) 506-3275
http://www.natoa.org

National Cable Television Association (NCTA)
1724 Massachusetts Avenue N.W.
Washington, D.C. 20036
TEL: (202) 775-3550
http://www.ncta.com

National Electrical Contractors Assoc. (NECA)
3 Bethesda Metro Center, Suite 1100
Bethesda, MD 20814
TEL: (301) 657-3110
http://www.necanet.org

National Electrical Manufacturers Association (NEMA)
1300 North 17th Street, Suite 1847
Rosslyn, VA 22209
TEL: (703) 841-3200
http://www.nema.org

National Institute of Standards Technology (NIST)
100 Bureau Drive, Stop 3460
Gaithersburg, MD 20899-3460
TEL: (301) 975-6478
http://www.nist.gov

National Research Council of Canada
Building M58, Montreal Road Ottawa
Canada K1A 0R6
TEL: (613) 998-7352
http://www.nrc.ca

National Systems Contractors Association (NSCA)
625 First Street SE, Suite 420
Cedar Rapids, IA 52401
TEL: (319) 366-6722
http://www.nsca.org

North American Association of Telecommunications Dealers (NATD)
1045 E. Atlantic Avenue, Suite 206
Delray Beach, FL 33483
TEL: (561) 266-9440
http://www.natd.com

OECC/IOOC
Level 4, 66 King Street
Sydney NSW 2000
Australia
TEL: 61-2-9262-2277
http://www.tourhosts.com.au/OECCIOOC/

Optoelectronic Industry and Technology Development Association (OITDA)
Sumitomo Edogawabashiekima Building
7F, 20-10 Sekiguci I-chome
Bunkyo-ku Tokyo, 112-0014, JAPAN
TEL: (81) 35225-6431
http://www.oitda.or.jp

Optical Society of America (OSA)
2010 Massachusetts Avenue N.W.
Washington, D.C. 20036
TEL: (202) 223-8130
http://www.osa.org

Reed Exhibition Companies
International Security Conference (East)
Jacob Javits Convention Center
New York, NY
International Security Conference (West)
Las Vegas Convention Center
Las Vegas, NV
http://www.iscreedexpo.com

Society for Broadcast Engineers (SBE)
9247 N. Meridian Street, Suite 305
Indianapolis, IN 46260
TEL: (317) 846-9000
http://www.sbe.org

Society of Cable Television Engineers (SCTE)
140 Philips Road
Exton, PA 19341
USA: (800) 542-5040
TEL: (610) 363-6888
http://www.scte.org

Society of Telecommunications Consultants (STC)
13766 Center Street, Suite 212
Carmel Valley, CA 93924
USA: (800) STC-7670
http://www.stcconsultants.org

The Information Technology & Telecommunications Association (TCA)
PO Box 278076
Sacramento, CA 95827-8076
http://www.tca.org

Telecommunications Industry Association (TIA)
2500 Wilson Blvd. St. 300
Arlington, VA 22201
TEL: (703) 907-7700
http://www.tiaonline.org

Trade Associates, Inc.
11820 Parklawn Drive
Rockville, MD 20852
TEL:(301) 468-3210

Underwriters Laboratories (UL)
333 Pfingsten Road
Northbrook, IL 60062-2096
TEL: (847) 272-8800
http://www.ul.com

U.S. Army Communications - Electronics Command (CECOM)
AMSELCOM D-4
Fort Monmouth, NJ 07703
TEL: (908) 544-3163

United States Telecom Association (USTA)
1401 H St. NW, Suite 600
Washington, D.C. 20005-2164
TEL: (202) 326-7300
http://www.usta.org

Wire Association International
1570 Boston Post Road
Guildford, CT 06437
TEL: (203) 453-2777
http://www.wirenet.org

Wire & Cable Industry Suppliers Association
3869 Darrow Road, Suite 109
Stow, OH 44224
TEL: (330) 686-9544
http://www.wiretech.com/wcisa/

FIBER OPTIC & RELATED TRADE MAGAZINES

Applied Optics
Optical Society of America
2010 Massachusetts Avenue, NW
Washington, D.C. 20036-1023

Broadband Networking News
PBI Media, LLC Corporate Office
1201 Seven Locks Road, Suite 300
Potomace, MD 20854
TEL: (301) 354-2000
http://www.pbimedia.com

Broadcast Engineering
Primedia Business Magazines & Media
PO Box 12914
Overland Park, KS 66282-2914

Broadband Guide
Pennwell
1421 S. Sheridan
Tulsa, OK 74112
USA: (800) 331-4463
http://www.broadband-guide.com

Cabling Installation & Maintenance
Ten Tara Boulevard
Fifth Floor
Nashua, NH 03062
http://www.broadband-guide.com/cim

CCTV Applications & Technology
Burke Publishing Company
15825 Shady Grove Road, Suite 130
Rockville, MD 20850

Communication News
2500 Tamiami Trail, North
Nokomis, FL 34725

Communications Convergence
CMP Media, Inc.
600 Community Drive
Manhasset, NY 11030
TEL: (516) 562-5000

Communications Week
CMP Publications, Inc.
600 Community Drive
Manhasset, NY 11030
TEL: (516) 562-5000

Communications Technology
PBI Media, LLC Corporate Office
1201 Seven Locks Road, Suite 300
Potomace, MD 20854
TEL: (301) 354-2000
http://www.pbimedia.com

ComNet
111 Speen Street
PO Box 9107
Framingham, MA 01701-9515

EDN
Cahners Business Information
8773 S. Ridgeline Boulevard
Highlands Ranch, CO 80126-2329

EE Product News
Penton Media, Inc.
1100 Superior Avenue
Cleveland, OH 44114-2543

Electro Optics
Milton Publishing Company Ltd.
5 Tranquil Passage, Blackheath, London SE3 OBY
ENGLAND

Electronic Buyer's News
CMP Media, Inc.
600 Community Drive
Manhasset, NY 11030
TEL: (516) 562-5000

Electronic Design
Penton Publishing Inc.
611 Route # 46 West
Hasbrouck Heights, NJ 07604

Electronic Engineering Times
CMP Media, Inc.
600 Community Drive
Manhasset, NY 11030
TEL: (516) 562-5000

Electronic Products

Hearst Business Communications
645 Stewart Avenue
Garden City, NY 11530
TEL: (516) 227-1300
http://www.electronicproducts.com

Fiber Optics Newsletter

Information Gatekeepers, Inc.
214 Harvard Avenue
Boston, MA 02134

Fiberoptic Product News

Cahners
301 Gibraltar Dr., Box 650
Morris Plains, NJ 07950
http://www.fpnmag.com

Fiber Optic News

PBI Media, LLC Corporate Office
1201 Seven Locks Road, Suite 300
Potomace, MD 20854
TEL: (301) 354-2000
http://www.pbimedia.com

Fibre Systems

Institute of Physics Publishing LTD
Dirac House, Temple Back
Bristol, BSI, 6BE, UK
TEL: +44 (0) 117-929-7481
http://www.fibre-systems.com

Guidelines

Corning Incorporated Optical Fiber
One Riverfront Plaza
Corning, NY 14831
http://www.corning.com/opticalfiber

Integrated Communications Design (ICD)

Pennwell
1421 S. Sheridan
Tulsa, OK 74112
USA: (800) 331-4463
http://www.icd.pennet.com

LAN TIMES

McGraw-Hill
7050 Union Park Center, Suite 240
Midvale, Utah 84047

Lasers & Optronics

301 Gibralter Drive
PO Box 650
Morris Plains, NJ 07950-0650
http://wwww.lasersoptrmag.com

Laser Focus World

Pennwell
1421 S. Sheridan
Tulsa, OK 74112
USA: (800) 331-4463
http://lfw.pennet.com/home.cfm

Lightwave

Pennwell
1421 S. Sheridan
Tulsa, OK 74112
USA: (800) 331-4463
http://.lw.pennet.com/home.cfm

Microwaves & RF

Penton Publishing, Inc.
1100 Superior Avenue
Cleveland, OH 44114
http://www.mwrf.com

Military & Aerospace Electronics

Pennwell
1421 S. Sheridan
Tulsa, OK 74112
USA: (800) 331-4463
http://www.milaero.com

Network Magazine

CMP Media, Inc.
600 Community Drive
Manhasset, NY 11030
TEL: (516) 562-5000
http://www.network-mag.com

Optics & Photonics News

Optical Society of America
2010 Massachusetts Ave., NW
Washington, D.C. 20036
http://www.osa.org

Outside Plant

2615 Three Oaks Road, Suite 1B
Cary, IL 60013
TEL: (847) 639-2200
http://www.ospmag.com

Photonics Spectra

Laurin Publishing Co.
Editorial Offices
Berkshire Common
PO Box 4949
Pittsfield, MA 01202-4949
http://www.photonicsspectra.com

Security Magazine

Business News Publishing
Security Group Home Office
1050 IL Route 83, Suite 200
Bensonville, IL 60106-1096
TEL: (603) 616-0200
http://www.securitymagazine.com

Security Management

1625 Prince Street
Alexandria, VA 22314
http://www.securitymanagement.com

Sound & Communications Magazine

25 Willowdale Avenue
Port Washington, NY 11050
TEL: (516) 767-2500
http://www.soundandcommunications.com

Sound & Video Contractor

Primedia Business Magazines & Media
PO Box 12914
Overland Park, KS 66282-2914

Spectrum Magazine — IEEE

Editorial Board
New York, NY 10001
TEL: (212) 419-7555
http://www.spectrum.ieee.org

Video Systems

Primedia Business Magazines & Media
PO Box 12914
Overland Park, KS 66282-2914
http://www.industryclick.com/magazine.asp

WDM Solutions

Pennwell
1421 S. Sheridan
Tulsa, OK 74112
USA: (800) 331-4463
http://www.wdm.pennet.com

INDEX